Grade 6

Discovery Education | SCIENCE TECHBOOK

California
Volume 2

Unit 3: Causes and Effects of Regional Climates
Unit 4: Our Changing Climate

Copyright © 2019 by Discovery Education, Inc. All rights reserved. No part of this work may be reproduced, distributed, or transmitted in any form or by any means, or stored in a retrieval or database system, without the prior written permission of Discovery Education, Inc.

NGSS is a registered trademark of Achieve. Neither Achieve nor the lead states and partners that developed the Next Generation Science Standards were involved in the production of this product, and do not endorse it.

To obtain permission(s) or for inquiries, submit a request to:
Discovery Education, Inc.
4350 Congress Street, Suite 700
Charlotte, NC 28209
800-323-9084
Education_Info@DiscoveryEd.com

ISBN 13: 978-1-68220-554-9

Printed in the United States of America.

1 2 3 4 5 LBC 28 27 26 25 24   C

Acknowledgments

Acknowledgment is given to photographers, artists, and agents for permission to feature their copyrighted material.

Cover and inside cover art: SusanGaryPhotography / Moment / Getty Images

Letter to the Student

Dear Student,

See science in a whole new way! In this class, you'll be using California Science Techbook™. California Science Techbook is a science program developed by Discovery Education. The program is full of images, videos, Hands-On Activities, digital tools, reading passages, animations, and other activities. These resources will help you analyze and interpret data. You will solve problems and make connections between science and the world around you. California Science Techbook is made so you can work at your own pace and explore questions you may have about science. You'll even be able to see your progress in real time using the online Student Dashboard.

You will use this Student Edition to explore important ideas and record what you know and what you have learned. You'll also use it to make connections to the digital content in online Techbook. This will help you develop your own scientific understanding.

In each section of the Student Edition, you'll find QR codes. When you scan these codes, they'll take you to the online Science Techbook section you need. For instance, QR links throughout the book take you directly to important anchor or investigative phenomena videos and images. Once you are inside digital California Science Techbook, you can try some Explorations, Hands-On Activities, or Virtual Labs. All of them will help you explore the most important ideas in a concept. Enjoy this deep dive into the exciting world of science!

Sincerely,

The Discovery Education Science Team

Letter to the Parent/Guardian

Dear Parent/Guardian,

This year, your student will be using California Science Techbook™, a comprehensive science program developed by the educators and designers at Discovery Education. California Science Techbook is an innovative program that offers engaging, real-world problems to help your student master key scientific concepts and act and think like a scientist. Students engage with interactive science instruction to analyze and interpret data, think critically, solve problems, and make connections across science disciplines. In addition, they experience dynamic content, explorations, videos, digital tools, Hands-On Activities and labs, and game-like activities that inspire and motivate scientific learning and curiosity.

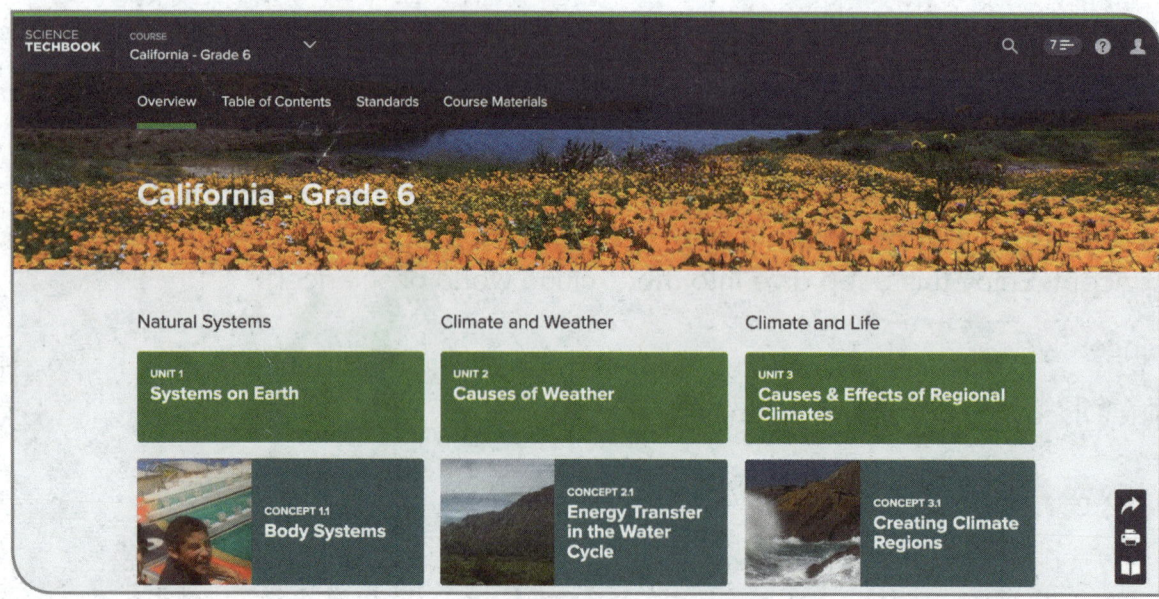

Within this Student Edition, you'll find QR codes that take you and your student to a corresponding section of Science Techbook online. Once in Techbook, students will have access to the Core Interactive Text of each concept, as well as thousands of resources and activities that build deep conceptual scientific understanding. Additionally, tools and features such as the Interactive Glossary and text-to-speech functionality allow Science Techbook to target learning for students of a variety of abilities.

To use the QR codes, you'll need to download a free QR reader. Readers are available for phones, tablets, laptops, desktops, and virtually any device in between. Most use the device's camera, but there are some that scan documents that are on your screen.

For resources in California Science Techbook, you'll need to sign in with your student's username and password the first time you access a QR code. After that, you won't need to sign in again, unless you log out or remain inactive for too long.

We encourage you to support your student in using California Science Techbook. May you and your student enjoy a fantastic year of science!

Sincerely,

The Discovery Education Science Team

Table of Contents

Unit 3: Causes and Effects of Regional Climates

Unit Overview
Anchor Phenomenon: Annual Average Temperature 2
Unit Project Preview: Engineering a Better Banana 4

Concept 3.1 Creating Climate Regions
Concept Overview 6
Engage 8
   Investigative Phenomenon: Deserts
Explore 20
Explain 54
Elaborate 58
Evaluate 62

Concept 3.2 Environmental and Genetic Influences
Concept Overview 64
Engage 66
   Investigative Phenomenon: Caribou on Thin Ice
Explore 76
Explain 100
Elaborate 102
Evaluate 106

Concept 3.3 Reproductive Success

Concept Overview 108
Engage 110
   Investigative Phenomenon: Nests and Hatchlings
Explore 118
Explain 146
Elaborate 150
Evaluate 156

Concept 3.4 Heredity

Concept Overview 158
Engage 160
   Investigative Phenomenon: Maintaining Potato Biodiversity
Explore 168
Explain 188
Elaborate 190
Evaluate 196

Unit Wrap-Up

Unit Project: Engineering a Better Banana 198

Unit 4: Our Changing Climate

Unit Overview
Anchor Phenomenon: Measuring Climate Change 206
Unit Project Preview: Cow Pollution 208

Concept 4.1 Causes of Climate Change
Concept Overview... 210
Engage... 212
   Investigative Phenomenon: Climate Change Models
Explore..222
Explain..262
Elaborate..266
Evaluate...270

Concept 4.2 Climate Change Impacts Organisms
Concept Overview...272
Engage...274
   Investigative Phenomenon: Migration Journey of the Monarch Butterfly
Explore..282
Explain..306
Elaborate.. 310
Evaluate...314

Concept 4.3 Reducing Human Impacts on the Environment

Concept Overview 316
Engage 318
   Investigative Phenomenon: Cattle Population in the United States
Explore 326
Explain 358
Elaborate 360
Evaluate 364

Unit Wrap-Up

Unit Project: Cow Pollution 366

Resources

Glossary R1
Index R20

UNIT 3

Causes and Effects of Regional Climates

Unit 3: Causes and Effects of Regional Climates

UNIT 3 | Get Started

Annual Average Temperature

The pattern is clear: it is warm in the tropics and cold at the poles. At the global scale, what is the reason this occurs? In this unit, you will learn about the causes for different climate regions and the impact that these regions have on living things.

Quick Code
ca6505s

Guiding Questions

1. Why is the climate so different in different regions of the planet?

2. Why are organisms so different in different regions of the planet?

3. What makes organisms so similar to but also different from their parents?

4. What makes animals behave the way they do, and how does their behavior affect their survival and reproduction?

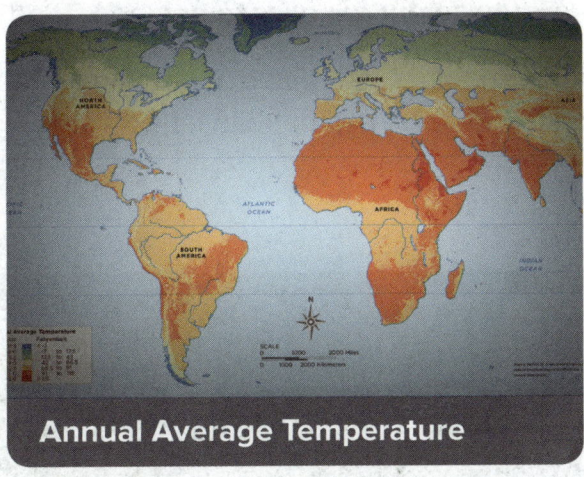

UNIT 3 | Unit Project Preview

Solve Problems

Unit Project: Engineering a Better Banana

Thinking About Solutions

Could we use what we know about genetics to engineer food?

You are part of a team of scientists interested in helping farmers optimize their crops. Before moving forward, brainstorm traits farmers may want to control and why.

Quick Code
ca6506s

Banana Tree

Use the graphic organizer to sort your ideas.

Trait 1

Trait 2

CONCEPT 3.1

Creating Climate Regions

Student Objectives

By the end of this lesson:

- [ ] I can develop and refine a model that describes and predicts how differences in solar energy received cause temperature variations at different locations on Earth.
- [ ] I can synthesize information to identify patterns in air and ocean currents in local and global environments.
- [ ] I can synthesize information to construct explanations for how Earth's shape and solar radiation cause Earth's climates.
- [ ] I can develop a model that describes and predicts how solar radiation and convection cause air and ocean currents.
- [ ] I can synthesize information to construct explanations for how energy transfer, Earth's rotation, winds, and ocean currents interact to produce Earth's climates.

Key Vocabulary

Antarctic, Arctic, climate, conduction, convection (weather), Coriolis effect, current, density, energy transfer, equator, global wind, heat, kinetic energy, latitude, light energy, ocean current, polar, radiation, salinity, solar energy, temperate, tropical, water cycle, water vapor, wind

Quick Code
ca6508s

Activity 1
Can You Explain?

Why do different places on Earth have different climates?

Quick Code
ca6509s

CONCEPT 3.1 | Why do different places on Earth have different climates?

Activity 2
Ask Questions

Deserts

Watch the video. As you watch, **note** what you see and **make predictions** about what might happen next.

Quick Code
ca6510s

Let's Investigate Deserts

Your Ideas

Describe what you see in the video and **make predictions** about what might happen next.

_____
_____
_____
_____
_____
_____
_____
_____

What questions do you have about deserts?

_____
_____
_____
_____
_____
_____
_____
_____

Concept 3.1: Creating Climate Regions

CONCEPT 3.1 | Why do different places on Earth have different climates?

Activity 3
Observe

Climate Zones

Watch the videos. **Answer** the questions that follow.

Quick Code
ca6511s

Climate

Climate Zones

CCC Stability and Change

Climate

Answer the questions based on the videos.

Are the differences in climate random, or do they follow a pattern? What factors do you think affect climate the most?

_____
_____
_____
_____
_____
_____
_____

Concept 3.1: Creating Climate Regions | 13

CONCEPT 3.1

Why do different places on Earth have different climates?

What is the difference between weather and climate?

_____
_____
_____
_____
_____
_____
_____

How are different climates described?

_____
_____
_____
_____
_____
_____
_____
_____

How does climate affect how people live?

CONCEPT 3.1

Why do different places on Earth have different climates?

Activity 4
Evaluate

What Do You Already Know About Creating Climate Regions?

Quick Code
ca6512s

Global Climate Zones

The following world map is divided into climate zones. The key shows which colors correspond to each climate zone.

Write the correct name of each climate zone in the box above the key:

| tropical | dry | temperate | continental | polar |

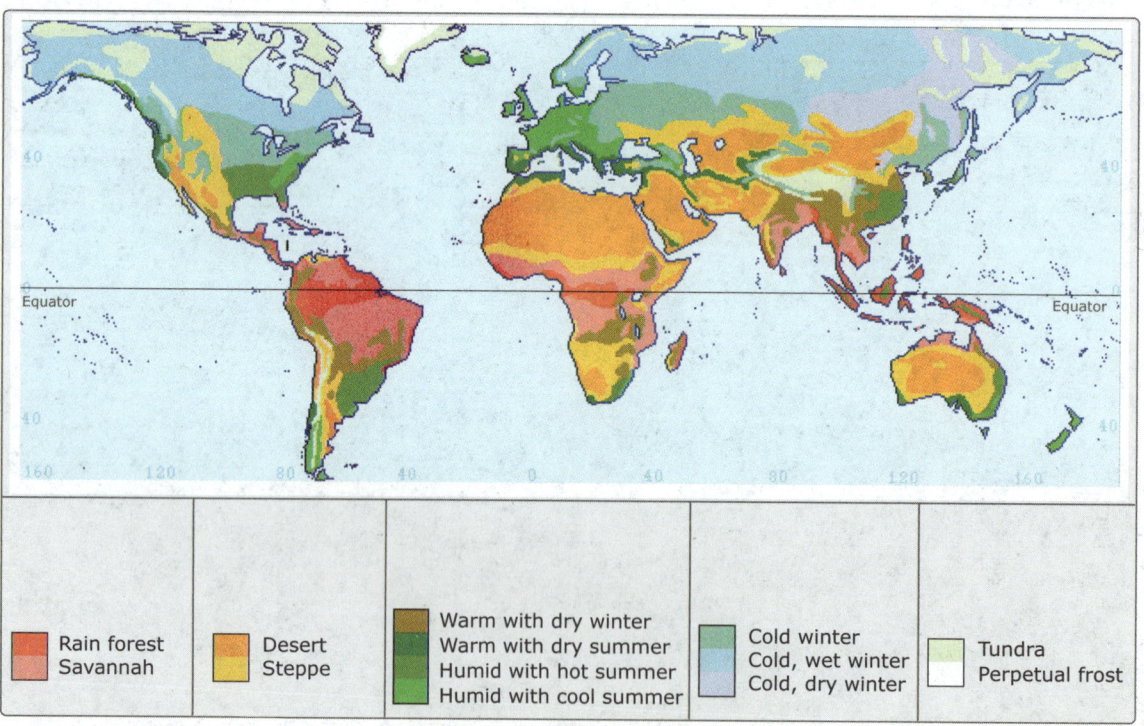

Climate Factors

Based on how the climate zones are arranged across the globe, what do you think is an important factor in determining climate?

_____

_____

_____

_____

_____

_____

_____

_____

_____

_____

_____

_____

_____

_____

_____

CONCEPT 3.1

Why do different places on Earth have different climates?

Earth's Tilt and Climate

This image shows Earth tilted on its axis. Among the four locations (labeled A, B, C, and D), which is receiving the most direct sunlight? **Circle** your answer.

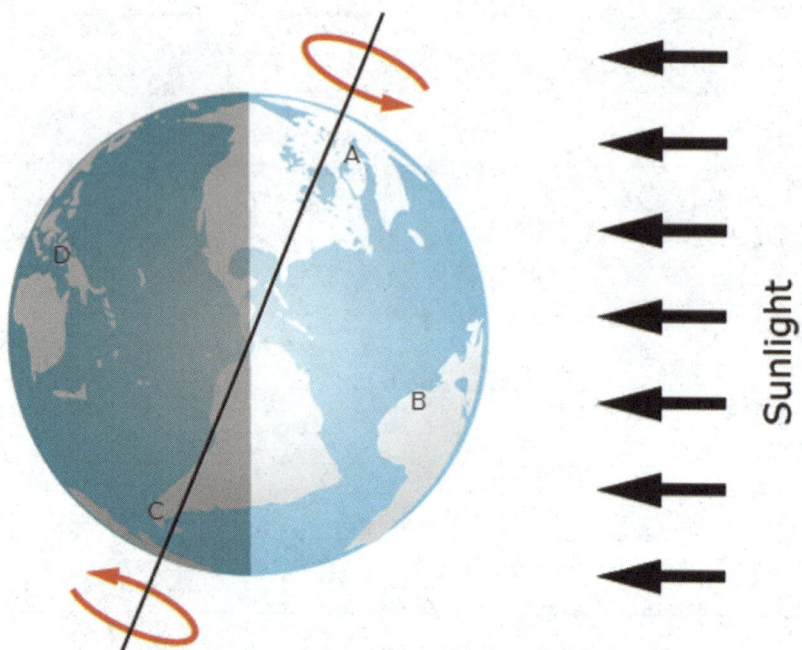

Consider the relationship between Earth's axial tilt and direct sunlight, explored in the previous item. What do these factors have to do with differences in Earth's climate?

_____
_____
_____
_____
_____
_____
_____
_____

Heating Water

In the following image, a pot of water is being heated by a flame. Think about how warmer and cooler regions of water would move around this pot as it is heated. Which sets of arrows show the correct movement of warm and cool water regions?

Circle the arrows that apply.

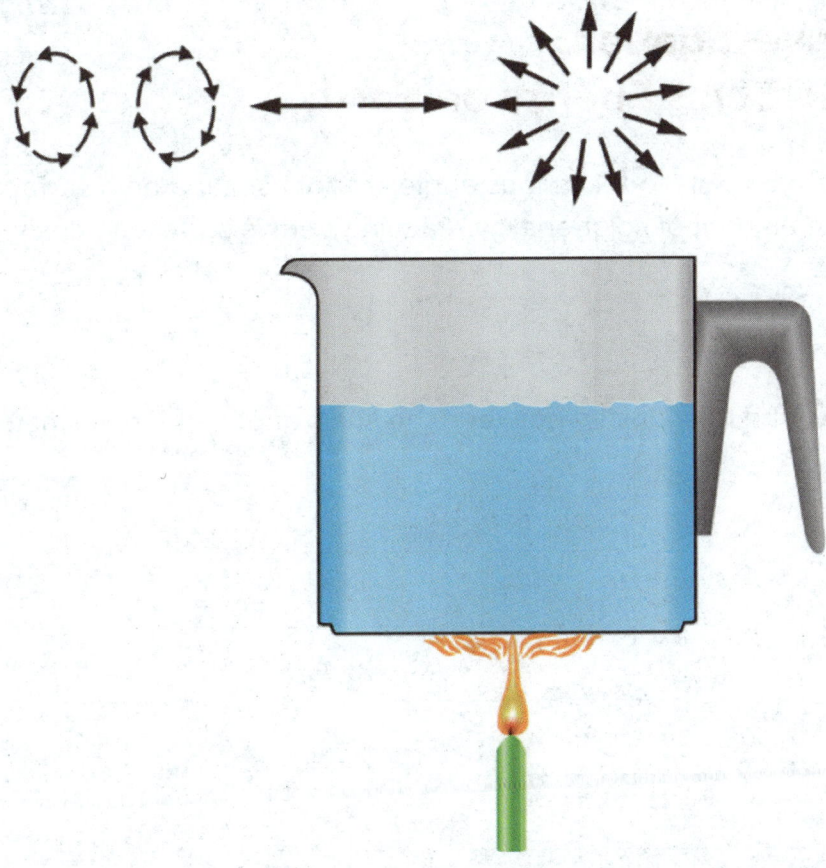

CONCEPT 3.1 — Why do different places on Earth have different climates?

What Is the Relationship among Earth's Shape, Latitude, and Climate Regions?

Activity 5
Investigate

Hands-On Investigation: Modeling the Sun's Energy on Earth

Quick Code
ca6517s

In this investigation, you will model various angles of light shining on a surface to investigate how the amount of solar energy reaching Earth's surface varies with latitude.

Predict

What do you predict the relationship between the sun's angle and the climate will be?

_____
_____
_____
_____
_____
_____
_____
_____
_____

SEP Developing and Using Models

What materials do you need? (per group)

- Graph paper
- Plastic chips
- Flashlight
- Batteries, size D
- Metric ruler

Procedure

Hold the flashlight above the graph paper in various positions. For each position, trace the outline of where the light falls on the page. Label each outline with the approximate angle you used to hold the flashlight.

What are the limitations of this model?

_____
_____
_____
_____

How might this model be modified to better represent Earth and the sun?

_____
_____
_____
_____
_____
_____
_____
_____

Concept 3.1: Creating Climate Regions | 21

CONCEPT 3.1 Why do different places on Earth have different climates?

Based in your observations, how might the angle of the sun's rays affect Earth's temperature?

_____
_____
_____
_____

Reflect

Based on your observations, what happens when you change the beam of light from a direct beam to an angled beam? Explain.

_____
_____
_____
_____
_____

What do these results indicate about the amount of energy radiating from the flashlight?

_____
_____
_____
_____
_____
_____

Earth's axis is tilted relative to its orbital path around the sun. What can you conclude about the sun's energy based on the flashlight angles?

Activity 6
Analyze

Heating of Earth and Climate Regions

Quick Code
ca6518s

Look at the map of Earth. For each climate you read about, **indicate** the location of the climate on the map. Include annotations describing the amount of direct sunlight and general temperatures for each location.

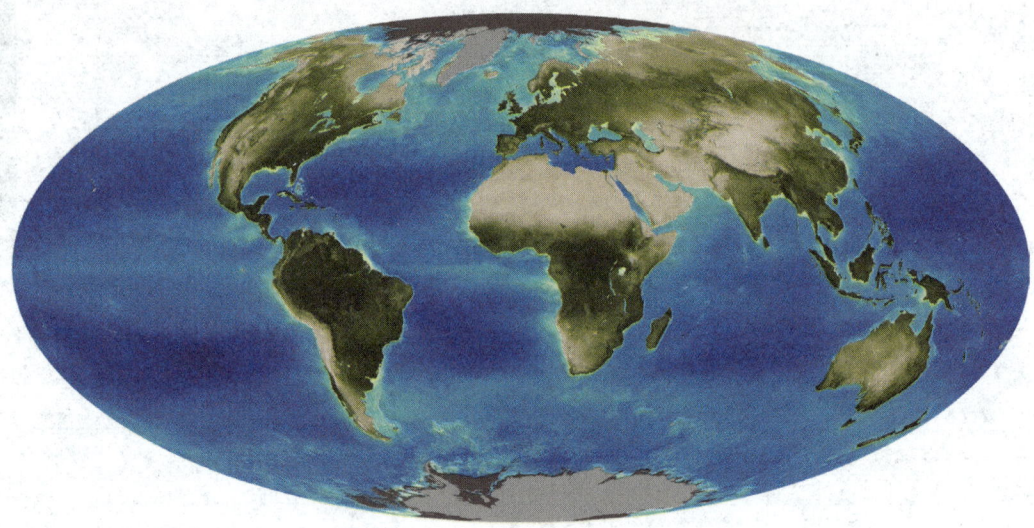

SEP — Obtaining, Evaluating, and Communicating Information

Heating of Earth and Climate Regions

Imagine you shine a narrow beam of light on a ball. If you shine the beam squarely onto the ball, it is a concentrated circle. If you shine it higher up on the ball, it spreads out into a larger elongated oval. The amount of energy radiating from the light does not change, but when the beam is shining on the edge of the ball, that energy is spread out over a larger area.

Earth is roughly spherical, and the amount of **solar energy** that reaches a given area of the surface decreases as you move away from the point at which the sun appears to be directly overhead. This means at higher latitudes, toward the poles, there is less solar energy falling on Earth per square meter than at lower latitudes. The amount of solar energy a region receives (per square meter) affects its **climate**.

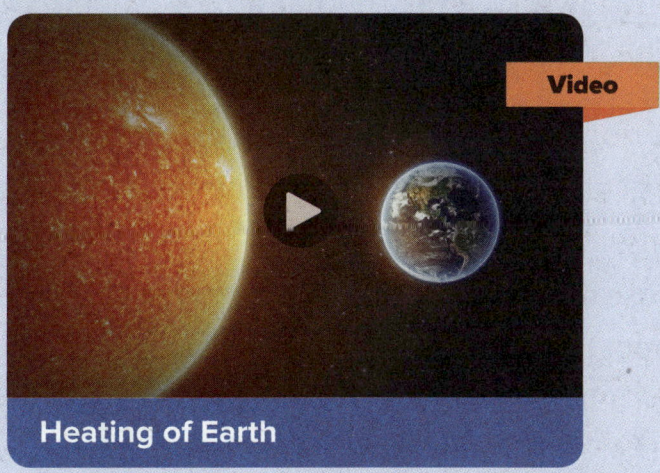

Heating of Earth

Heating of Earth and Climate Regions *cont'd*

Earth's climate regions can be grouped into three major zones: **polar**, **temperate**, and **tropical**. These three zones are determined primarily by **latitude**, which also affects temperature.

The **tropical** zone is located along the **equator** roughly between the Tropic of Cancer (23.5° north) and the Tropic of Capricorn (23.5° south). Because the equatorial regions receive the most direct sunlight, temperatures are warm year-round. In some places in the tropical zone, the humidity is high. Although most **tropical** regions do not experience seasons like many parts of the United States do, some experience dry and wet seasons known as monsoons. Precipitation is very heavy during monsoon season.

The **polar** zones are found around the North and South Poles, from each pole at 90° roughly to the **Arctic** and **Antarctic** Circles at 66.5° latitude. There, sunlight hits Earth's surface at a low angle, giving this zone Earth's lowest average temperatures. Polar ice caps remain frozen throughout the year; however, polar regions do have seasons.

The temperate zones are located between the tropical and polar zones. These regions get less direct sunlight throughout the year than the tropical regions, but more than the polar regions. Temperatures are in between those of the tropical and polar zones, and generally vary more throughout the year. These areas experience seasonal changes in temperature and precipitation, and many experience snow during the winter. Most of the United States is in the temperate zone. There is much variety in rainfall in the temperate zone, including areas with high, medium, and low precipitation.

Activity 7
Evaluate

Solar Energy Distribution and Mixed Messages

Quick Code
ca6519s

Solar Energy Distribution

Construct an explanation about how Earth's shape influences climates found at different regions. Use the model you created with graph paper and a flashlight to **support** your explanation.

_____
_____
_____
_____
_____
_____
_____
_____
_____
_____
_____
_____
_____

SEP Constructing Explanations and Designing Solutions

Why do different places on Earth have different climates?

Mixed Messages

Below are a few sentences about the global climate zones you just studied. Revise these sentences so that they make more sense.

> The most direct sunlight strikes the temperate zone. This is the main reason why it is consistently the coldest region on Earth. Less direct sunlight strikes the polar zones. These zones have warmer and colder seasons because of Earth's tilt. The least amount of direct sunlight reaches the tropical zones. These regions stay warm year-round.

What Is the Relationship among Wind, Energy Transfer, and Climate?

Activity 8
Observe

Quick Code
ca6520s

Traveling Rubber Ducks

Rubber ducks spilled into the ocean in 1992 and traveled great distances over 15 years. **Watch** the video and **answer** the questions that follow.

Traveling Rubber Ducks

What patterns in the wind and ocean currents took the plastic toys for a ride?

_____
_____
_____
_____

Concept 3.1: Creating Climate Regions | 29

CONCEPT 3.1 Why do different places on Earth have different climates?

Write something you found interesting about the topic.

_____
_____
_____
_____

Describe something that caused you to say "oh!"

_____
_____
_____
_____

Write a question about something you learned or want to learn more about.

_____
_____
_____
_____

Activity 9
Analyze

Convection within Earth's Systems

What is the relationship among wind, temperature, and climate? Read the text and watch the videos. **Look** for evidence to answer the question.

Quick Code
ca6521s

Convection within Earth's Systems

Imagine you are an oceanographer studying ocean surface currents. You learn about several containers full of bath toys that washed overboard in a storm in the north Pacific Ocean. You know that the waves will carry the toys away from the site of the shipwreck in the ocean. An oceanographer would ask questions about the **wind** and ocean currents, and the role that **energy transfer** plays to predict where these bath toys would end up. What will the **climate** be at their destinations?

Every day, energy from the sun arrives on Earth. Despite the sun's distance of 150 million kilometers from Earth, light and **heat** energy from the sun travel to our planet. Without this energy, life as we know it would not exist. If the sun is so far away, how can its heat and **light energy** travel to Earth without being lost or transferred along the way? The answer is that sunlight travels by **radiation**.

Radiation is the process of heat transfer through electromagnetic waves. Electromagnetic waves can interact with different types of matter.

Concept 3.1: Creating Climate Regions | 31

Convection within Earth's Systems *cont'd*

For example, the microwaves in a microwave oven will cause water molecules to vibrate very quickly. This causes an increase in temperature. This increase in temperature is going to be very important in helping us predict how those bath toys travel around the oceans.

Earth has a global wind system that consists of winds that blow in a constant direction over long periods of time. The planet is circled by six wind belts, three on either side of the **equator**. Their direction is determined by the amount of solar radiation received at different latitudes and the rotation of Earth.

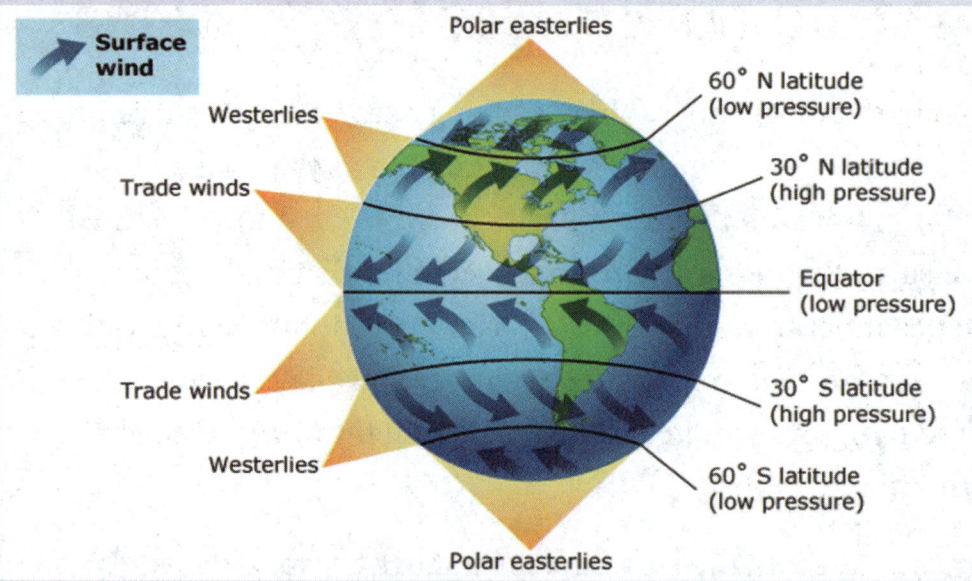

Air warmed by the sun's radiation rises. The warm air spreads out and becomes less dense. This creates an area of low air pressure. The warm air is replaced by cooler, more densely packed air flowing from nearby. This process causes wind, and this type of energy transfer is called convection. Convection is a process of heat transfer in which warmer and cooler fluids or gases, such as water or air, move due to differences in **density**. Wind moves from areas of high pressure to low pressure as a result of convection.

Recall that solar radiation is most direct at the equator; the energy causes air to rise around the globe at this **latitude**. If the warm air contains enough **water vapor**, as the air rises it loses this water in the form of rain. Meanwhile, cooler air masses from north and south of the equator flow in to take the place of the rising warm air. As the warm air flows away from the equator, it cools and descends; by the time it reaches Earth's surface again, the air is dry. This dry air forms a band of deserts around the planet along 30° north and south latitude. The air then flows back toward the equator.

Earth's Wind

These processes produce a worldwide wind pattern in which surface winds move from about 30° latitude toward the equator. However, these global winds are also affected by the rotation of Earth on its axis. This is called the **Coriolis effect**. As surface air travels over Earth toward the equator, Earth's rotation causes the air to veer to the west. These surface winds are known as the trade winds. A similar circulation occurs between the poles and 60° latitude.

Coriolis Effect

Concept 3.1: Creating Climate Regions | 33

Convection within Earth's Systems *cont'd*

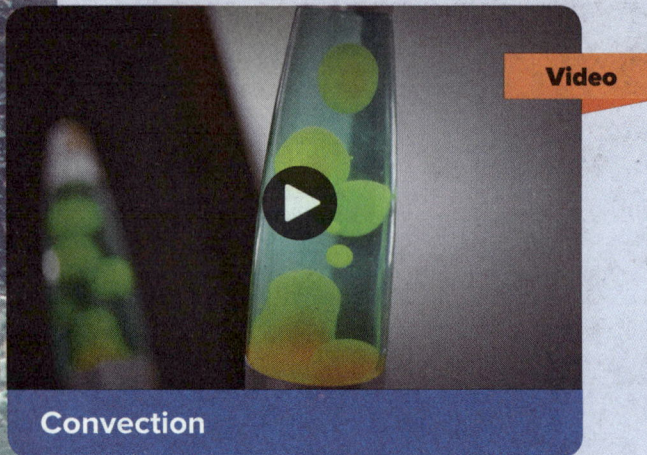
Convection

Another circulation, which causes surface winds to come from the west in the Northern Hemisphere, occurs between 60° and 30° latitude. Each of these wind currents also drives water currents in the oceans beneath them.

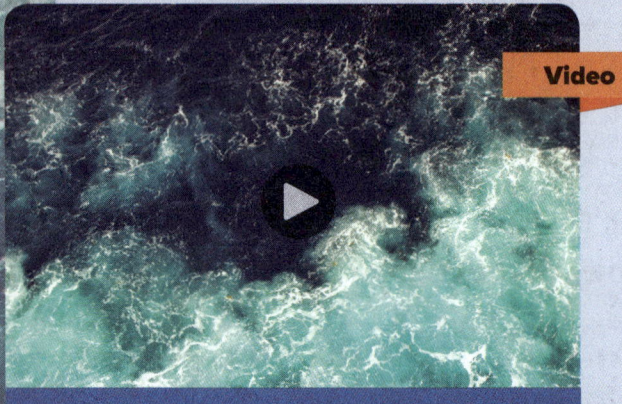
Surface Ocean Currents

The **global wind** system—and the resulting surface currents in Earth's oceans—move thermal energy and moisture around the planet. The wind system has a significant effect on the **water cycle**, and consequently, on weather and climate. For example, in the United States, the prevailing winds are from the west because the latitude is between 30° and 60°. These winds travel over the Pacific Ocean, gathering moisture until they reach coastal mountain ranges. Here, the air must rise to get over the mountains. As it rises, it cools, and the water vapor it contains condenses and falls as rain or snow. In another example, the westerly winds over the Atlantic Ocean drive warm water in ocean currents across the Atlantic. These movements produce the moderate temperatures and abundant rainfall in Great Britain.

Uneven heating of Earth creates global wind systems. How does that impact climate?

_____
_____
_____
_____
_____
_____

Earth's rotation causes the Coriolis effect. What impact does the Coriolis effect have on global winds?

_____
_____
_____
_____
_____
_____
_____

What role does wind play in creating surface ocean currents?

Activity 10
Evaluate

The Trade Winds

The diagram shows the trade winds, which are the predominant winds at the equator. **Examine** the diagram closely, and then **circle** the correct word or phrase to complete each statement.

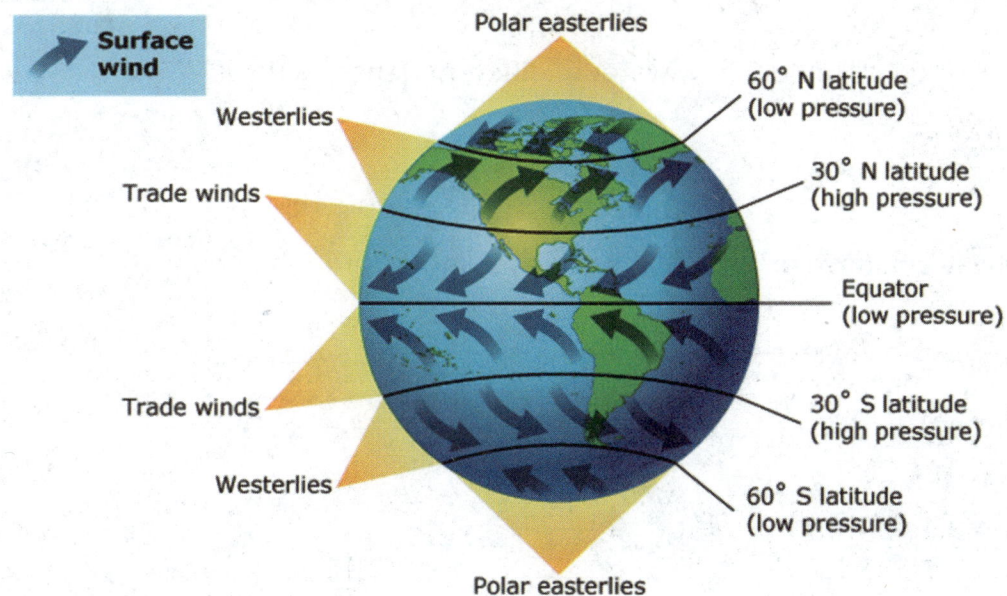

The equator receives more solar radiation than higher latitudes. This causes air at the equator to become **cooler and rise | cooler and fall | warmer and rise | warmer and fall.**

This movement of air creates an area of **low | high** pressure at the equator.

To keep the system in balance, air at mid-latitudes **falls, increasing pressure | falls, decreasing pressure | rises, increasing pressure | rises, decreasing pressure.**

Wind is created when air flows from areas of pressure **low to high | high to low.**

Concept 3.1: Creating Climate Regions | 37

CONCEPT 3.1 | Why do different places on Earth have different climates?

What Is the Relationship among Ocean Currents, Energy Transfer, and Climate?

Activity 11
Investigate

Hands-On Investigation: Density Currents

Quick Code
ca6523s

In this activity, you will model how water of different densities flows through Earth's oceans.

Predict

How does water behave at different densities to move through oceans?

_____
_____
_____
_____
_____
_____
_____
_____

 SEP Planning and Carrying Out Investigations

 SEP Constructing Explanations and Designing Solutions

What materials do you need? (per group)

- Beakers, 250 mL
- Graduated cylinder, 250x2 mL
- Balance, triple beam
- Spoon, plastic
- Food coloring
- Wax pencil
- Salt
- Water
- Bowls, plastic
- Safety goggles (per student)
- Lab apron (per student)
- Gloves (per student)

Procedure

Plan at least four tests: two tests will involve combining water with different amounts of salt, and two will involve combining water of different temperatures. You can see the currents that form in these different situations by using the beakers, food coloring, and different temperatures of water. Outline your proposed experimental design on the provided lines and submit it to your teacher for approval.

CONCEPT 3.1 — Why do different places on Earth have different climates?

Use the provided table to record your results.

	Salt Amount 1	Salt Amount 2	Temperature 1	Temperature 2
Written observation				
Illustrated observation				

Reflect

Based on your observations, which types of water are denser? Explain.

_____
_____
_____
_____
_____
_____
_____
_____

Where on Earth do these types of waters exist?

_____
_____
_____
_____
_____
_____
_____

What type of movement is shown by the water? Explain.

_____
_____
_____
_____
_____
_____
_____
_____
_____

Describe how the water movement you observed in this activity is similar to density currents in the oceans. How do you think these types of ocean waters help density currents form in the oceans?

What do you think would happen if you combined the cold, salty water with the warm, salty water in this experiment?

Concept 3.1: Creating Climate Regions

Activity 12
Analyze

Explanation of Density Currents

Quick Code
ca6524s

Examine the following images. Based on the images, how do you think ocean currents drive climate?

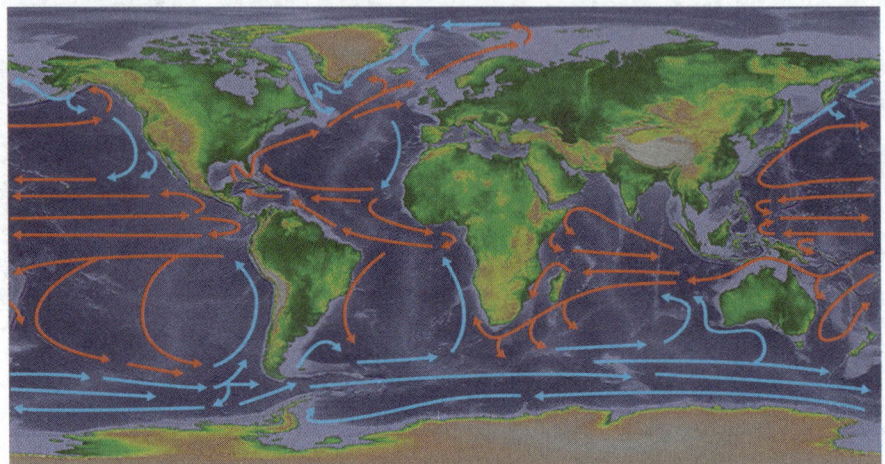

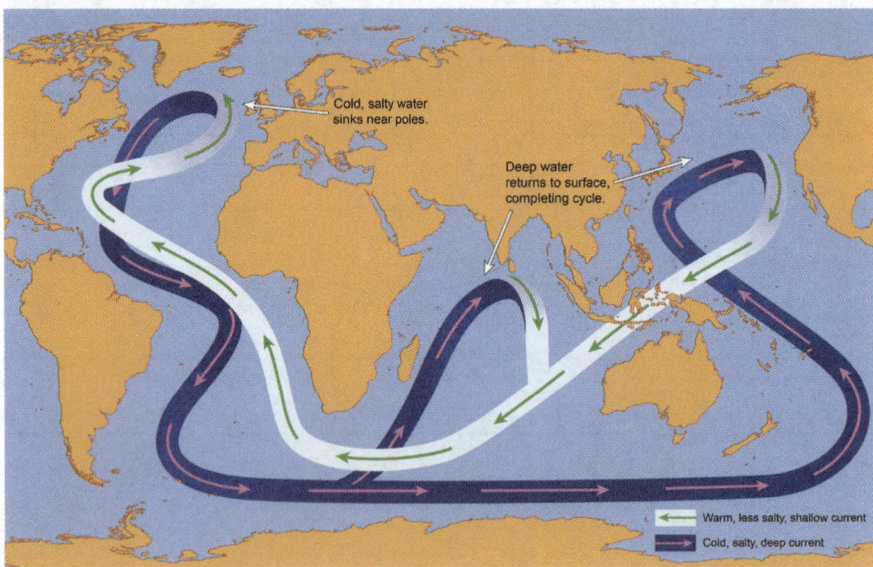

Now, read the text about density currents. Then, **review** the images that follow and **write** your own captions based on what you have learned from the text.

Explanation of Density Currents

Earth's ocean waters move in a continuous flow around the planet. These flowing waters are called currents. Some currents move at the surface of the oceans. These are called surface currents. Some currents move deeper underwater, or even along the ocean floor. These are called **density** currents.

Surface currents in the ocean tend to follow the path of global winds. As winds blow across the ocean's surface, the **wind** closest to the water surface creates friction between the air and the water. Through this friction, surface water is pulled in the same direction as the wind. Just as global winds curve as they blow across Earth, ocean currents curve too. For example, the Gulf Stream is a **current** that flows northeast from Florida, then turns eastward into the Atlantic Ocean. The Gulf Stream is one of many currents of water in the Atlantic Ocean that all generally move in a clockwise direction. Because it carries a flow of warm water, the Gulf Stream is responsible for helping to warm the climates of the land masses that it passes.

Ocean currents are also affected by other factors. The movement of tectonic plates can change currents because the shape of an ocean's basin affects how ocean water flows. Similarly, tidal patterns caused by gravitational pull affect ocean currents. Water density differences related to **salinity** and water temperature are other significant factors in how currents move through the ocean.

Density currents form due to differences in ocean salinity and temperature. Salinity is a measure of how much salt is dissolved

Explanation of Density Currents *cont'd*

in ocean water. Generally speaking, salinity is fairly constant throughout the oceans. However, differences in temperature across the planet can change ocean salinity levels and cause ocean water to sink and rise.

For example, near the North and South Poles, temperatures are so low that the surface ocean water freezes. When water freezes, salt is excluded because there is no space for it in the ice that forms. The ice that forms is therefore made of freshwater. The excluded salt dissolves in any unfrozen water nearby, which then contains more salt per volume of water than it did before. This makes the water denser, and the saltwater sinks toward the ocean floor.

Scientists use the term *global conveyor belt* to describe the cycle of ocean currents around the planet. Cold, dense saltwater sinks to the bottom of the ocean floor. It flows toward the **equator**, where warmer air temperatures and increased **solar energy** gradually warm the water.

Recall that solar **radiation** is stronger near the equator. In addition to radiation, **heat** can also be transferred by **conduction**. Conduction is the process of heat transfer in which the particles in substances directly transfer **kinetic energy** to one another. When water is warmed by the sun, it causes the particles to move more quickly. When these energetic particles strike the particles in a colder region, they transfer some of their kinetic energy directly to them. This makes the particles in the colder region move faster, which increases the temperature of the water in that region. Because warm water is

46

less dense than cold water, the water begins to rise back toward the surface as it reaches the equator. Thus, conduction and convection work together. Along the way, warm water mixes with less salty water and becomes less dense. This warm, less-dense water then flows back toward the poles, driven by surface winds, and the cycle begins again. Water that moves as a density current can take hundreds of years to move through one entire cycle.

Now, **look** again at the images you examined at the beginning of the activity. Use what you learned from the text to **write** a title and caption for each image.

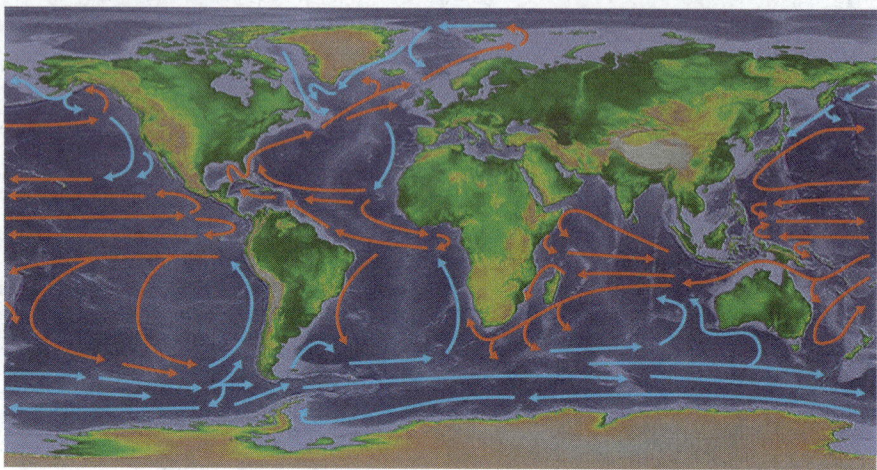

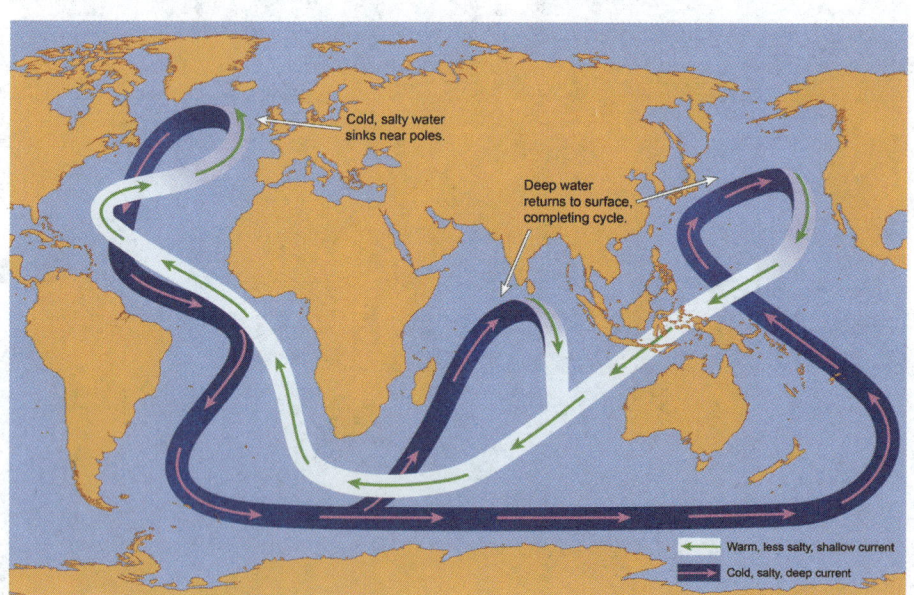

Activity 13
Observe

What Determines Climate?

Go online to explore how a region's climate is influenced by its elevation, proximity to the equator, and warm or cold ocean currents.

Quick Code
ca6525s

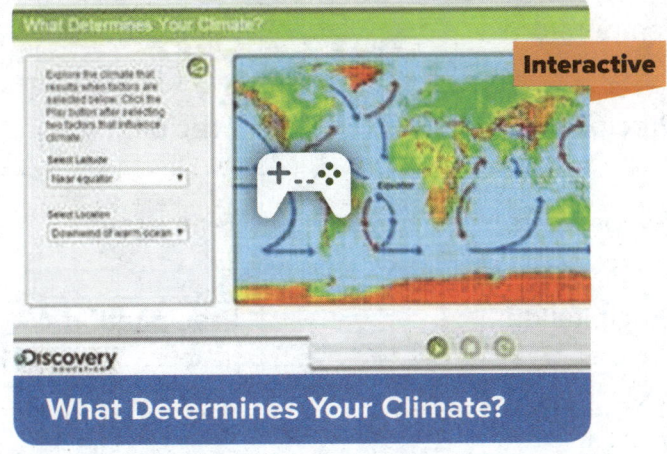

CONCEPT 3.1 — Why do different places on Earth have different climates?

Collect temperature and precipitation data in the chart below.

	Stats	Near Equator	Stats	Far from Equator
Downwind of warm ocean	Temperature		Temperature	
	Precipitation		Precipitation	
Downwind of cold ocean	Temperature		Temperature	
	Precipitation		Precipitation	
Far from ocean	Temperature		Temperature	
	Precipitation		Precipitation	
High elevation	Temperature		Temperature	
	Precipitation		Precipitation	

How are these patterns related to the pattern of particle movement of the air and water?

_____
_____
_____
_____
_____

Activity 14
Evaluate

Climate Currents

The diagram shows the climate zones of the United States. Use information from the diagram and what you have learned about how ocean currents affect climate to **answer** the questions that follow.

Quick Code
ca6526s

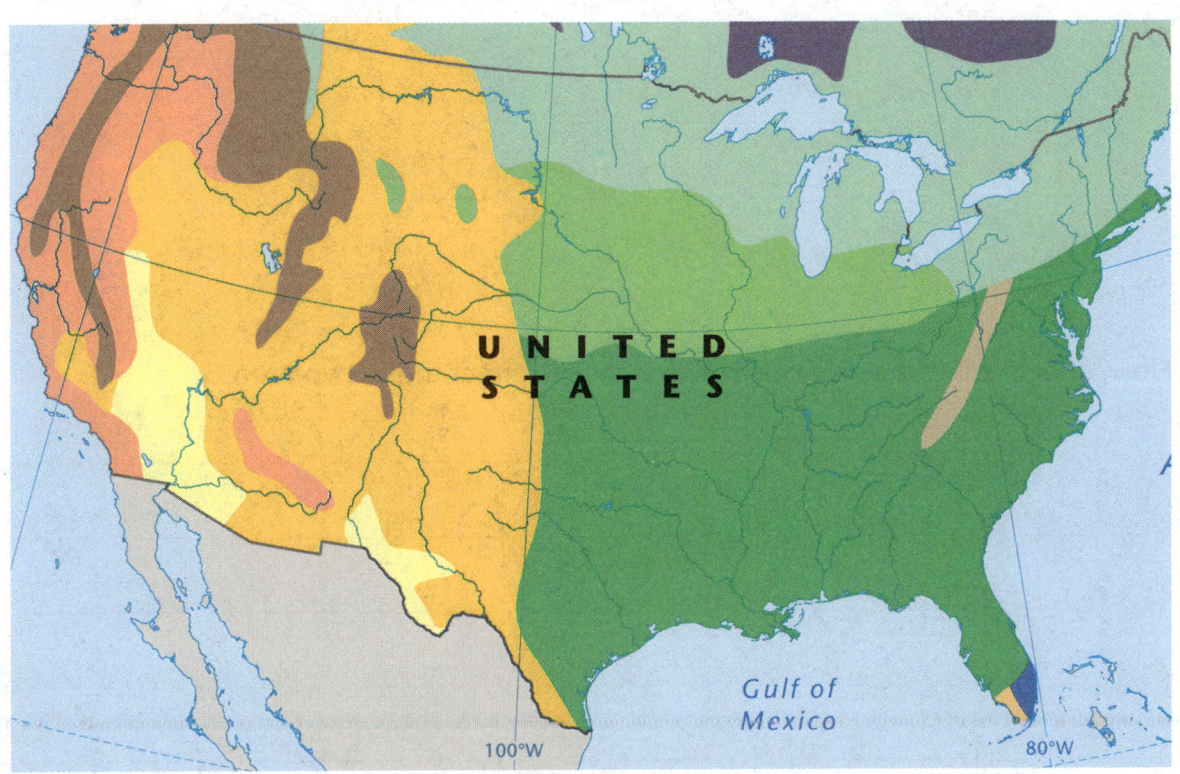

Concept 3.1: Creating Climate Regions | 51

CONCEPT 3.1 Why do different places on Earth have different climates?

What are two separate areas of the United States that share the same climate?

How, specifically, do you think ocean currents affect these two areas?

EXPLORE

Concept 3.1: Creating Climate Regions | 53

CONCEPT 3.1 | Why do different places on Earth have different climates?

Activity 15
Record Evidence

Deserts

Quick Code
ca6527s

As you worked through this lesson, you investigated and gathered evidence about creating climate regions. Now, take another look at the Deserts video, which you first saw in Engage.

How has your understanding of the Deserts video changed?

Read the Can You Explain? question from the beginning of this lesson.

 Can You Explain?

Why do different places on Earth have different climates?

Use your new understanding of the Deserts video to write a scientific explanation answering a question. To help you formulate your scientific explanation, fill in the table to organize your ideas. Recall that a scientific explanation contains three elements: a scientific claim, evidence to support the claim, and reasoning that connects the evidence to the claim. Be sure to include evidence that describes qualitative or quantitative data that you gathered to support your claim.

SEP Constructing Explanations and Designing Solutions

Scientific Claim

Evidence to Support Claim

Reasoning That Connects Evidence to Claim

Concept 3.1: Creating Climate Regions

CONCEPT 3.1 | Why do different places on Earth have different climates?

Write your scientific explanation in the spaces provided.

STEM in Action

Activity 16
Analyze

Inside the Ocean's Currents

Quick Code
ca6528s

Read the text and **watch** the video. As you read and watch, **think** about how the ocean currents compare in different parts of the world.

Inside the Ocean's Currents

Ocean currents have a significant effect on Earth's **climate**, but currents are nearly impossible to track simply by watching the movement of water. Some currents follow **wind** patterns over thousands of miles, while others flow deep underneath the ocean's surface. Scientists who study ocean currents have to be creative in how they track and measure currents.

Tracing the Movement of Water Masses

Scientists have many methods for studying the flow of ocean currents. Sometimes marine biologists attach tracers onto aquatic animals to track their migration along the "highways" of the ocean. These busy highways are strong ocean currents that can help carry animals across very long distances. For migratory animals, traveling with the flow of the ocean saves precious energy.

Going with the Flow

Use this map of ocean currents to **answer** the following questions. Work with a small group to write your answers.

Pacific Currents

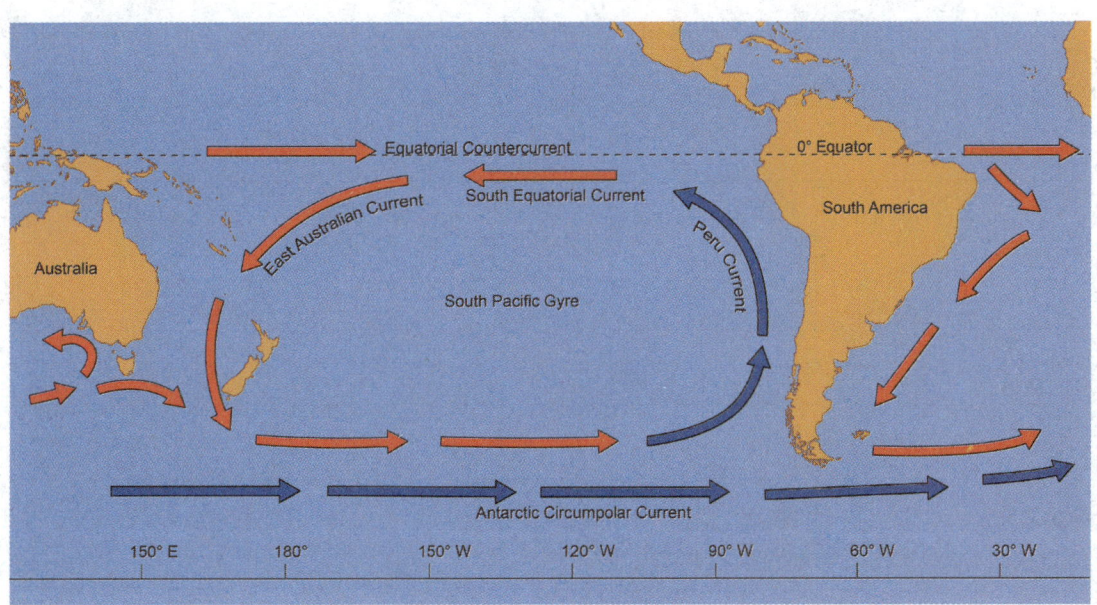

SEP Obtaining, Evaluating, and Communicating Information

How does the South Equatorial Current flow compare to the flow of the Equatorial Countercurrent?

If red arrows indicate warm water and blue arrows cool water, then what kind of water makes up the Peru Current? How do you know?

CONCEPT 3.1
Why do different places on Earth have different climates?

Activity 17
Concept Review

Review: Creating Climate Regions

Quick Code
ca6529s

Now that you have completed the objectives for this concept, review the core ideas you have learned. Record some of the core ideas below.

Core Ideas

_____
_____
_____
_____
_____
_____
_____
_____
_____
_____
_____
_____
_____

Talk with a Group

Now, think about the annual average temperature map you saw in Get Started. Discuss how what you've learned about creating climate regions can help you understand the annual average temperature map.

CONCEPT

3.2

Environmental and Genetic Influences

Student Objectives

By the end of this lesson:

- [ ] I can argue from evidence that living things are well-adapted to the climates in which they live.
- [ ] I can synthesize information from many ecosystems to describe and predict patterns in adaptations of living things.
- [ ] I can investigate the effect of various abiotic factors on a plant's growth and analyze data to evaluate the significance of each factor.
- [ ] I can synthesize information to predict how both environmental and genetic factors affect an organism's growth.

Key Vocabulary

adaptation, biome, desert, ecosystem, latitude, offspring, rain forest, taiga, tundra

Quick Code
ca6531s

Activity 1
Can You Explain?

How do the environment and genetics influence the growth of caribou in the Arctic?

_____
_____
_____
_____
_____
_____
_____
_____
_____

Quick Code
ca6532s

Concept 3.2: Environmental and Genetic Influences

CONCEPT 3.2
How do the environment and genetics influence the growth of caribou in the Arctic?

Activity 2
Ask Questions

Caribou on Thin Ice

Quick Code ca6533s

Watch the following video about the migration of caribou in the Arctic biome.

Let's Investigate Caribou on Thin Ice

What questions do you have about caribou migration?

___

___

___

Activity 3
Analyze

Arctic Tundra

View the image, read the text, and watch the video to answer the question: What features of the environment influence why caribou migrate?

Quick Code
ca6534s

ENGAGE

Arctic Tundra

The Arctic **tundra** is one of Earth's terrestrial, or land, biomes. Located near Earth's poles, the Arctic tundra **biome** is characterized by having a very cold climate, little rainfall, and a short growing season. There is a permanently frozen layer called the permafrost underneath thin topsoil. Although this climate is extremely cold, many plants and animals call this biome home. In addition to the caribou, polar bears, wolves, foxes, hares, and lemmings are just a few of the animals in the Arctic tundra. And while you will not find trees in the tundra, there are low-lying shrubs, flowering plants, lichens, mosses, and grasses.

Migrating Caribou

CCC Cause and Effect

Arctic Tundra *cont'd*

Within each biome, there are many ecosystems—interactions between living and nonliving things. In an **ecosystem**, the plants and animals and other living things all rely upon the nonliving things, like water, soil and air, for survival.

The caribou migrate between different ecosystems in the summer and winter.

Arctic Tundra Biome Facts: Its Climate and Geography

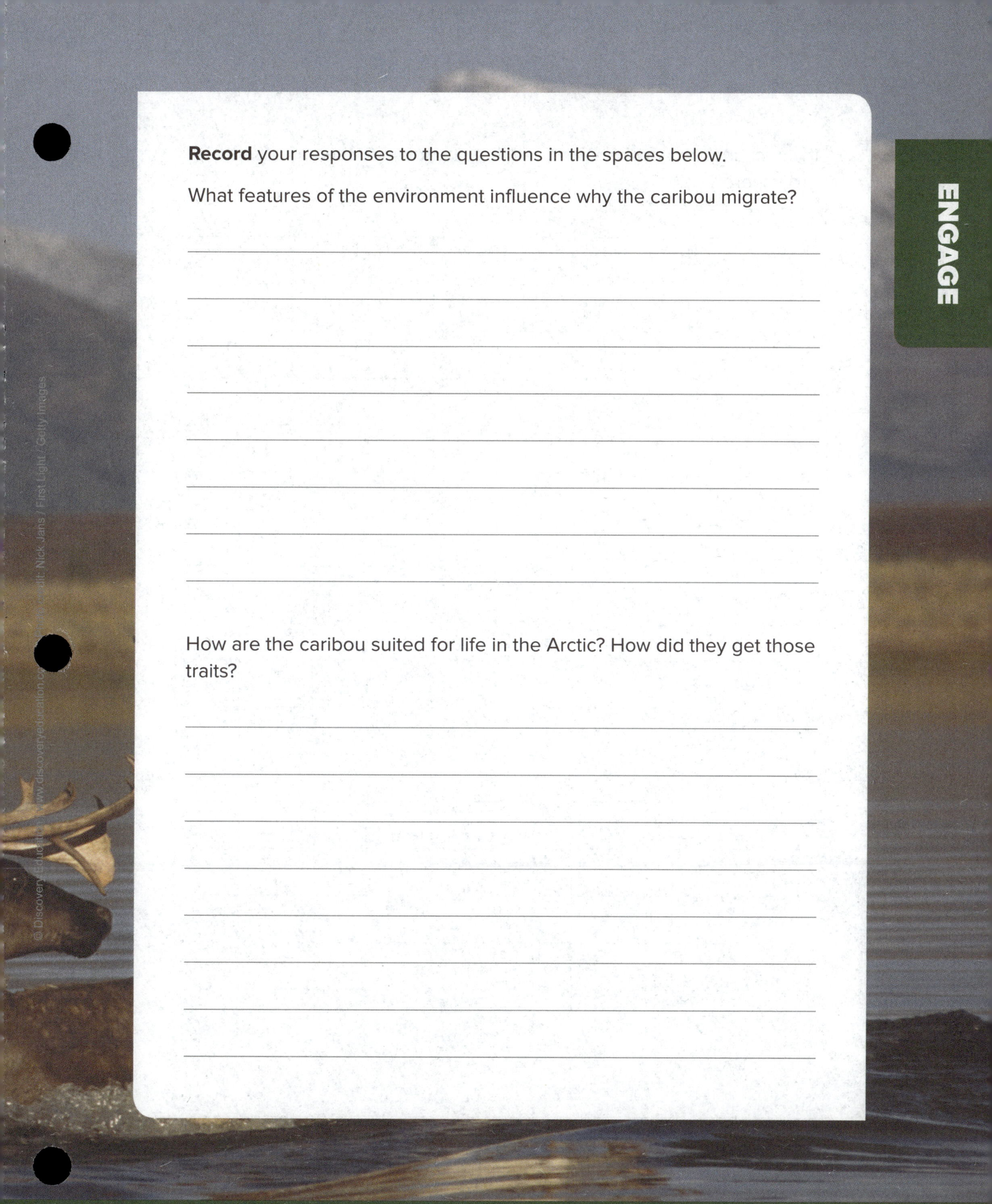

Record your responses to the questions in the spaces below.

What features of the environment influence why the caribou migrate?

_____

_____

_____

_____

_____

_____

How are the caribou suited for life in the Arctic? How did they get those traits?

_____

_____

_____

_____

_____

_____

_____

_____

If you change the environment, will that change how the organisms grow there?

Activity 4
Evaluate

What Do You Already Know About Environmental and Genetic Influences?

Quick Code
ca6535s

Basic Needs

What are an organism's basic needs for survival? What happens when the organism struggles to meet those needs?

_____
_____
_____
_____
_____
_____
_____
_____
_____
_____
_____
_____

Concept 3.2 — How do the environment and genetics influence the growth of caribou in the Arctic?

Influencing Growth

Write the number of each environmental factor below next to the type of organism behavior it would most likely affect in the table. Some environmental factors may be matched to more than one type of behavior.

1. Availability of light
2. Water
3. Size of habitat

Environmental Factor(s)	Organism Behavior Affected
	a plant that carries out photosynthesis
	trees and shrubs growing in an ecosystem
	the amount of food that an animal decides to forage for
	animal growth from a baby to an adult
	number of different species living in one place

Inherited Traits

Select all statements that indicate that a genetic factor has most likely influenced the growth of living things.

- [ ] A puppy becomes a full-size adult just like his dad.
- [ ] A rabbit has brown spots on its fur like its mom.
- [ ] A plant experiences dryness after record temperatures during August.
- [ ] For years, the grass plants are shorter than the tall flowering plants in a forest.
- [ ] Organisms compete for survival and a suitable habitat in an ecosystem.

CONCEPT 3.2 How do the environment and genetics influence the growth of caribou in the Arctic?

What Traits Are Beneficial in Specific Climate Regions?

Activity 5
Reason

Characteristics for the Climate

In this investigation, you will create a climate map that shows different climate zones in the United States. You will use this model along with your observations of organisms in your schoolyard to determine how climate influences the traits of living things.

Quick Code
ca6538s

What materials do you need? (per group)

- Outline map of the United States
- Map of climate zones in the United States
- Colored pencils

Procedure

1. In pairs, shade the map of the United States that you received with different colors, corresponding to different climate zones. Use the map of climate zones that occur in the United States as reference. Add labels to describe details.

2. Use your finished map to identify the climate zone where your school is located.

3. Then, go outside and explore the schoolyard. Look for animals and/or plants that are characteristic of your climate zone. Make a list of these organisms and choose one of them to collect as a sample.

4. Back in the classroom, share your sample with the class and explain why it is characteristic of your climate zone.

SEP Asking Questions and Defining Problems

Reflect

Why were the samples you collected representative of living things?

CONCEPT 3.2 | How do the environment and genetics influence the growth of caribou in the Arctic?

Why is it important to build the climate map before collecting samples of living things?

According to your map, how might your sample of a living thing survive in the climate zone to which it belongs?

Activity 6
Analyze

Traits for the Environment

Quick Code ca6539s

Why are polar bears white and grizzly bears brown in color? **Read** the text, **watch** the video, and **look** for evidence to support an answer to the question.

Traits for the Environment

Different kinds of environments have plants and animals with traits that benefit their survival. For example, the emperor penguin has thick blubber to keep it from freezing in the Antarctic. However, the Galapagos penguin, which lives near the equator, has a small, streamlined body that cools off fast in the hot locale. Even plants have beneficial characteristics. For instance, the cone of a spruce tree is shaped to allow it to shed ice and sleet in the harsh winters of the taiga. Having these traits helps them to grow and survive in their particular environment. How much influence does the environment play on an organism's growth? If the environment changes, does it impact the plants and animals living there?

Video — Animal Adaptations

Many animals have beneficial traits that help them survive their environment. How do these traits develop in organisms?

CONCEPT 3.2 | How do the environment and genetics influence the growth of caribou in the Arctic?

Activity 7
Evaluate

Where Are These Adaptations Useful?

Quick Code ca6540s

Use this assignment to think about how adaptations are specific to different types of environments.

Look at the animals below. **Think** about what adaptations they have and how those adaptations are useful to particular environments. **Write** next to each animal whether it most likely lives in the Arctic, a cave, a desert, or a tropical rain forest.

Animals	Environments
Poison Dart Frog	
Lizard	
Loon	

SEP Constructing Explanations and Designing Solutions

Salamander	
Cricket	
Fox	
Tree Frog	

How Do Environmental Factors Influence the Growth of Living Things?

Activity 8
Analyze

Nonliving Factors of Ecosystems

Quick Code
ca6541s

Read the text below. **Highlight** the nonliving factors that influence the survival of organisms in an ecosystem.

Nonliving Factors of Ecosystems

Plants and animals living in the same area rely upon each other to live and reproduce. This area is called an **ecosystem**. An ecosystem can be small, like a patch of open land between buildings that contains grass, insects, and dandelions. Or it can be quite large, like the Arctic, where caribou feed off the grasses and lichen, and wolves hunt the caribou and other prey. Regardless of the size of ecosystem, the organisms in an ecosystem depend upon environmental factors. These nonliving factors, such as sunlight, water, temperature, and amount of space, can influence the growth and survival of the organisms in the ecosystem.

You know that plants need water, sunlight, and air to grow. But did you know that the amount of light and the intensity of the light also will affect a plant's growth? In general, more intense light increases plant growth. However, light that is too intense can damage plant parts and cause drying or burning. Plants also respond to the amount of light and dark they receive daily. Some plants grow flowers when the days are longer than the nights. Some are the opposite. It is no wonder many people decorate during the fall with chrysanthemums. This plant only flowers when the days are shorter and the nights are longer.

Activity 9
Observe

One Tomato, Two Tomato

Go online to **design** an investigation to determine how different environmental factors, such as amount of water, fertilizer, and pH of soil, affect tomato growth.

Quick Code
ca6542s

One Tomato, Two Tomato

Concept 3.2: Environmental and Genetic Influences | 85

CONCEPT 3.2 — How do the environment and genetics influence the growth of caribou in the Arctic?

As you design your investigation, **record** your data in the chart.

Trial #	Soil Type	Peat Moss	Soil pH	Fertilizer	Water	Tomato Weight	Description of Plant
1							
2							
3							
4							
5							
6							
7							
8							
9							
10							
11							
12							
13							
14							
15							
16							
17							
18							
19							
20							

Activity 10
Interpret Data

Plant Height and Color

Quick Code
ca6543s

Read the following passage and use the graph of plant growth to **answer** the questions.

Plant Height and Color

You know that plants need water, sunlight, and air to grow. But did you know that the amount of light and the intensity of the light also will affect its growth? In general, more intense light increases plant growth. However, light that is too intense can damage plant parts and cause drying or burning. Plants also respond to the amount of light and dark they receive daily. Some plants grow flowers when the days are longer than the nights. Some are the opposite. It is no wonder many people decorate during the fall with chrysanthemums. This plant only flowers when the days are shorter and the nights are longer.

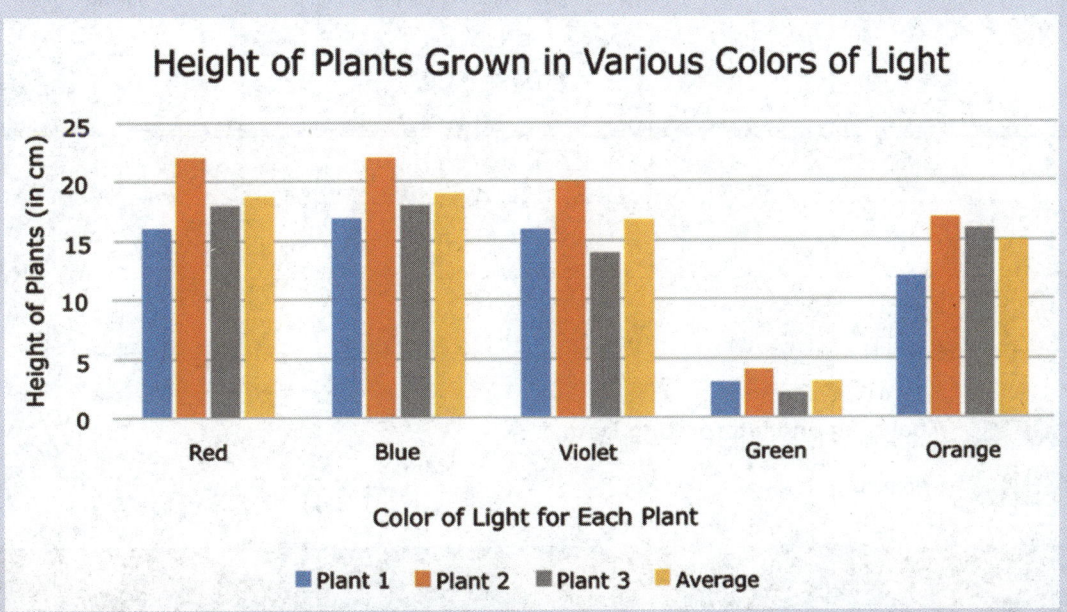

Plant Growth Graph

EXPLORE

Concept 3.2: Environmental and Genetic Influences

What patterns do you notice in the data?

Based on this data, can you support the claim that plants grown under violet lights grow at a significantly slower rate than plants grown under blue lights?

SEP Analyzing and Interpreting Data
CCC Patterns

Activity 11
Observe

How Does Light Affect Plant Growth?

Watch this video. Decide if you should change your answer to the questions in the previous activity. If yes, how?

Quick Code
ca6544s

How Does Light Affect Plant Growth?

Write your ideas here.

_____
_____
_____
_____
_____
_____
_____

Concept 3.2: Environmental and Genetic Influences

Concept 3.2

How do the environment and genetics influence the growth of caribou in the Arctic?

Activity 12
Evaluate

Analyze Plant Growth

Quinn investigated how fertilizer would affect the growth of radish plants. He grew five radish plants, each in its own pot. He added a different concentration of fertilizer to each pot. After several days of growth, he collected the following data:

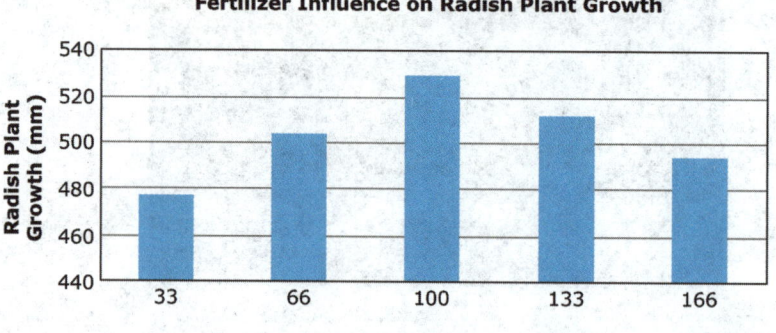

Fertilizer Influence on Radish Plant Growth

What should Quinn conclude about his results? **Select** the correct answer.

○ When more fertilizer is used, radish plants grow at a faster rate.

○ Radish plants grow the tallest when fertilizer concentration is 100 percent.

○ Radish plants will grow very tall in the absence of a fertilizer.

○ When the fertilizer concentration is 166 percent, radish plants are the shortest.

How Do Genetic Factors Influence Organism Growth?

Activity 13
Analyze

Dog Breeds

Quick Code
ca6546s

Read the text below and use the images to **identify** the inherited traits of the dog and plant species that can be observed.

Dog Breeds

Chihuahua and Great Dane

Dog breeds come in many different sizes. Despite their vast differences, all of our pet dogs, from the tiniest Chihuahua to the largest Great Dane, are the same species, called *Canis familiaris*. A Great Dane can stand an average of 30 inches tall, while a Chihuahua comes in at only about 8 inches. A Great Dane inherits its large size from its parents, just as the Chihuahua inherits its diminutive size from its parents. While each dog may not end up being the same exact size as its parents, you are never going to see a Chihuahua that is as tall as a Great Dane. The Chihuahua does not have the same genetic factors for height. A genetic factor controls what traits get passed down, or inherited, from parent to **offspring**.

Dog Breeds *cont'd*

Many ecosystems have a variety of plants that are different colors, shapes, and sizes. The short grasses in the Arctic tundra can be interspersed with an occasional bush or flowering plant. All of these plants are different heights even though they grow in the same soil and are exposed to the same amount of light.

Plants in the Arctic Tundra

How can this be? All living things inherit traits that influence how they grow. So, each of the plants in this ecosystem has different genetic factors that help determine how tall it can grow.

Species	Inherited Traits
Canis familiaris	
Plants in the Arctic tundra	

Concept 3.2: Environmental and Genetic Influences

Activity 14
Analyze

Getting to Know: Factors That Influence Human Growth and Development

Quick Code
ca6547s

Read the text. Then, **write** 3 Truths and 1 Lie about the factors that influence human growth.

Getting to Know: Factors That Influence Human Growth and Development

If you flip on the TV, chances are you will soon see a commercial that promises to "help you grow healthy and strong." If you believe everything you see on TV, you might think that all it takes to be healthy is some good cereal or fancy juice! Our diet can indeed affect how we grow and develop, but it is not the only factor involved in this process. In this lesson, you will explore some of the things that can impact the ways we grow and develop.

I thought people just grow naturally. Can outside factors affect how we grow?

Growing and developing are natural processes. Babies do not stay babies; they eventually become kids, then teenagers, then adults. However, there are some things that can affect how we grow. Many of the lifestyle choices we make can impact our health. You probably already know of some things that are bad for you or could harm your development. Smoking, drinking, and drugs can all have a very negative effect on our health. Other factors, like diet and exercise, also play a role in growth and development. Our bodies need proper nutrition, so a diet full of chips and soda will harm your growth. They may taste good at the time, but too many fatty foods will lead to obesity. Scientists estimate that approximately 30 percent of adults in the United States struggle with obesity, which leads to heart disease, high blood pressure, diabetes, and stroke.

Are there other outside factors that can impact how we grow?

EXPLORE

Concept 3.2: Environmental and Genetic Influences

Getting to Know: Factors That Influence Human Growth and Development *cont'd*

Yes, there are other outside factors, besides lifestyle choices, that we might not have the ability to control. We call these environmental factors. As you probably know, we all live in an environment. It is all of our surroundings, including our homes, schools, neighborhoods, and places of employment. If the environment around you is healthy and clean, chances are you do not need to worry. In some places, people lack many of the basic needs of life. In these places, health care might not be available. Children may not have access to routine immunizations. Work conditions might be unsafe, and water may be far away or unsafe to drink. Food can be hard to come by, and sanitation can be lacking, which can lead to illness. These are all environmental factors, and they can impact the way people grow and develop. You might live near a factory that pumps pollution into your air or a landfill that leaches toxic chemicals into your water. You might live in an area where the local stores do not sell fresh fruits and vegetables or where it is not safe to run and play outside. Unlike lifestyle choices, people usually do not have a lot of ability to control the environmental factors that are present in their lives.

I guess the only choice then is to move away, but then you have to have money. Are there any other factors that affect human growth and development?

Yes. There is a third set of factors that influences growth and development. Have you ever noticed how people in your family look similar? They might have the same type of hair or similar facial structure. This is called inheritance; we inherit genes and traits from our parents. Your parents each pass on genes to you. It is these genes that determine the way your earlobes hang, the length of your fingers, and how tall you can become with proper nutrition, as well as many other factors.

In addition to giving you positive traits, the genes from your parents may also carry genetic disorders. Some genetic disorders are caused by a mutation in a single gene. Others are caused by abnormalities within an entire chromosome, which contains many genes. Still others are caused by many different factors, including outside factors (heart disease, for example, can be caused by both genetics and lifestyle choices).

Truths	Lie

EXPLORE

CONCEPT 3.2 — How do the environment and genetics influence the growth of caribou in the Arctic?

Activity 15
Evaluate

Experimental Design

Quick Code ca6548s

A scientist ran a study to observe the growth of Ponderosa pine seedlings in different states. Plant growth was determined by measuring the height of seedlings that were 3 to 5 years old. The scientist collected the following data.

State Sampling Site	Average Plant Height (feet)
Montana	1.4
New Mexico	1.3
Colorado	1.7
South Dakota	1.9
Nebraska	1.6

CCC Cause and Effect

The scientist is interested in determining whether the differences in height are due to environmental or genetic factors. Help the scientist design this experiment by **putting the steps in the correct order.**

- Determine the average daily temperature and plant height at each sampling site.
- Observe differences in plant quality by looking at the quantity of needles on the tree.
- Compare plant height, daily temperature, and plant quality in each state.
- Collect plant seeds from each state sampling site.
- Plant and grow them in a state that is different from its original site.

Step	Steps for Designing the Experiment
1.	
2.	
3.	
4.	
5.	

Concept 3.2: Environmental and Genetic Influences

CONCEPT 3.2 | How do the environment and genetics influence the growth of caribou in the Arctic?

Activity 16
Record Evidence

Caribou on Thin Ice

As you worked through this lesson, you investigated and gathered evidence about environmental and genetic influences. Now, take another look at the Caribou on Thin Ice video, which you first saw in Engage.

Quick Code
ca6549s

Let's Investigate Caribou on Thin Ice

How has your understanding of the Caribou on Thin Ice video changed?

Read the Can You Explain? question from the beginning of this lesson.

> 💬 **Can You Explain?**
>
> How do the environment and genetics influence the growth of caribou in the Arctic?

 SEP Constructing Explanations and Designing Solutions

Use your new understanding of the Caribou on Thin Ice video to **write** a scientific explanation answering a question. Recall that a scientific explanation contains three elements: a scientific claim, evidence to support the claim, and reasoning that connects the evidence to the claim. Be sure to include evidence that describes qualitative or quantitative data that you gathered to support your claim.

Write your scientific explanation in the space provided.

Rebuilding a Community with Plants

Quick Code
ca6550s

Read the following text and **watch** the video to design your own community garden.

Rebuilding a Community with Plants

Did you know you could grow a garden to help people in your community? What environmental factors should you consider before creating this garden? How would genetic traits or environmental factors affect what you plant? Community gardens are a great way to give back. They help supply food to people who have limited access to fresh fruits and vegetables. They benefit the environment and create a space where people can work together for a great purpose.

Natasha Nicholes is a blogger who spends her time running the Union Avenue Community Garden in Chicago, Illinois. She realized there was an unmet need in her community. People did not have access to fresh fruits and vegetables. With this program, she can help meet this need by growing a garden.

The garden not only supplies food to people in the community but also allows people to help maintain the garden. This community garden also highlights the importance of using resources from the environment to survive.

Plants are living things. From sunlight to the amount of moisture in the air, many factors influence plant growth. This plant growth can directly affect just how well a garden turns out. Can you think of one change in the environment that might limit the number of vegetables or fruit produced for use?

A Community Garden

Identifying Factors

Read the factors listed in the word bank. **Circle** the environmental factor(s) that should be considered before growing a community garden.

plant color	shape of plant leaves
sunlight	different species of plants to plant
size of garden	soil moisture
water availability	

SEP Obtaining, Evaluating, and Communicating Information.

Garden Modeling

Purpose of your community garden

Design for your community garden

How you will keep it alive over time

ELABORATE

CONCEPT 3.2
How do the environment and genetics influence the growth of caribou in the Arctic?

Activity 18
Concept Review

Review: Environmental and Genetic Influences

Quick Code
ca6552s

Now that you have completed the objectives for this concept, review the core ideas you have learned. Record some of the core ideas below.

Core Ideas

Talk with a Group

Now, think about the annual average temperatures map you saw in Get Started. Discuss how what you've learned about environmental and genetic influences can help you understand the annual average temperatures map.

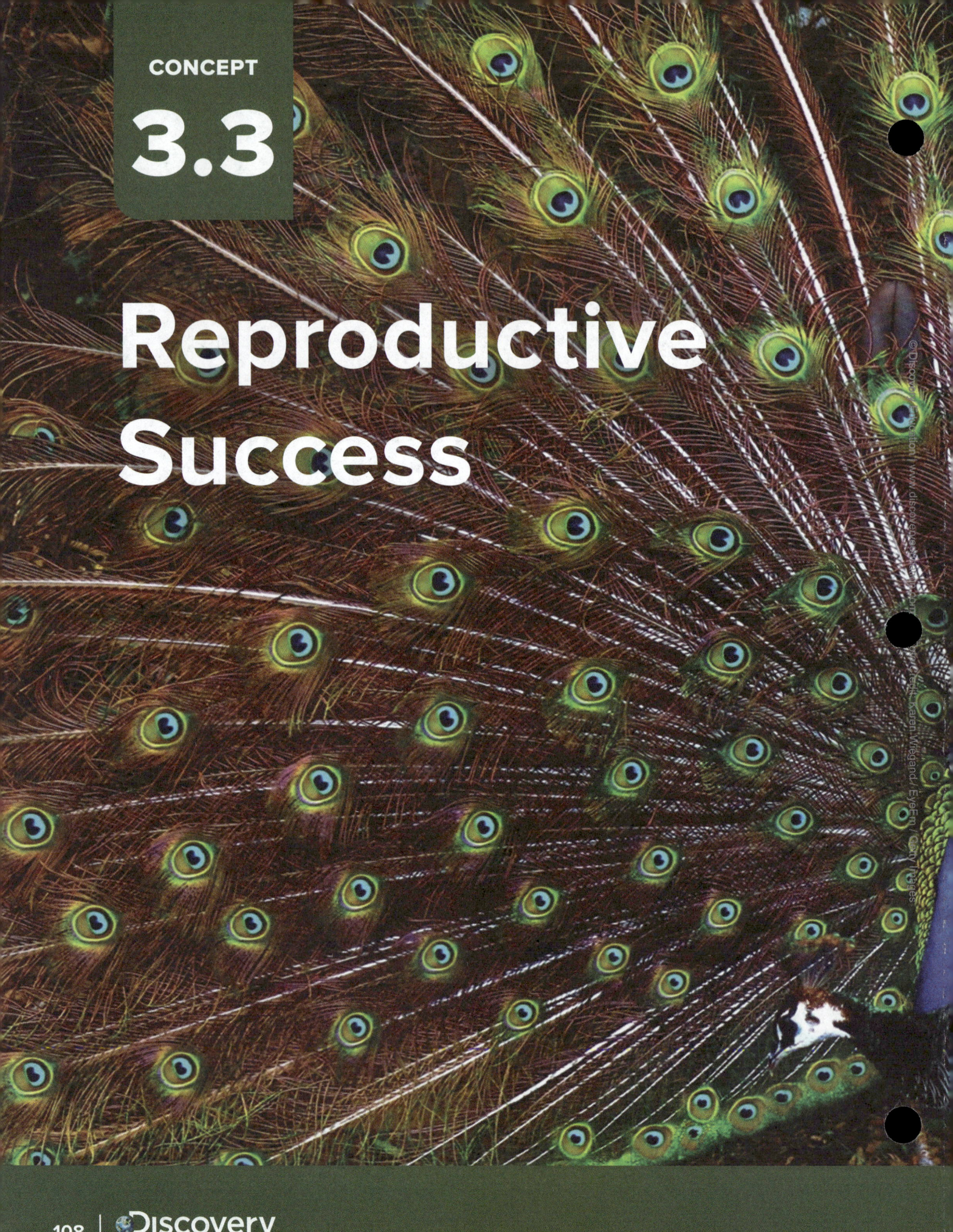

CONCEPT
3.3
Reproductive Success

Student Objectives

By the end of this lesson:

☐ I can develop models that describe the different patterns that result in the next generation of organisms produced by asexual reproduction and sexual reproduction.

☐ I can argue from evidence as to why sexual reproduction causes genetic variation among offspring.

☐ I can argue from evidence to support explanations that animal behaviors and specialized plant structures affect the probability of successful reproduction.

Key Vocabulary

asexual reproduction, behavior, egg, fertilization, gametes, instinct, offspring, pollination, reproduce, seed, sexual reproduction, sperm

Quick Code
ca6554s

Concept 3.3: Reproductive Success | 109

Activity 1
Can You Explain?

Are there certain climate conditions where asexual reproduction or sexual reproduction might be favored?

_____
_____
_____
_____
_____
_____
_____
_____
_____

Quick Code
ca6555s

CONCEPT 3.3 Are there certain climate conditions where asexual reproduction or sexual reproduction might be favored?

Activity 2
Ask Questions

Nests and Hatchlings

Watch the video and then **create** a social media post describing the video.

Quick Code
ca6556s

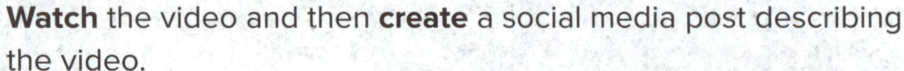

Let's Investigate Nests and Hatchlings

SEP Obtaining, Evaluating, and Communicating Information
CCC Structure and Function

Share with your social followers what you learned from the video. If you were going to include a picture in your social media post, what would you include?

What questions do you have about nests and hatchlings? Write them here.

_____
_____
_____
_____
_____
_____

ENGAGE

Concept 3.3: Reproductive Success | 113

| CONCEPT 3.3 | Are there certain climate conditions where asexual reproduction or sexual reproduction might be favored? |

Activity 3
Evaluate

What Do You Already Know About Reproductive Success?

Which Organisms Reproduce Asexually?

Circle the organisms that you think reproduce asexually or have the potential to do so.

Quick Code
ca6557s

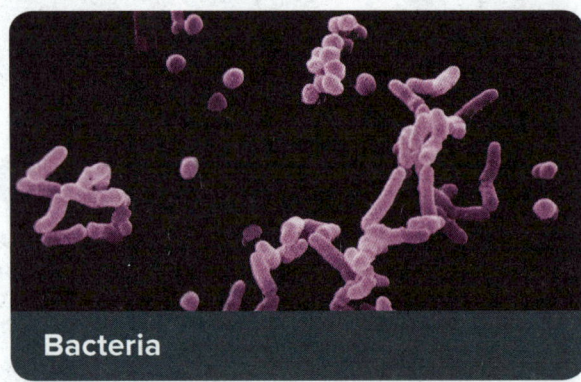

Bacteria

Snake

Cat

Moss

Potatoes

Bee

Concept 3.3: Reproductive Success | 115

CONCEPT 3.3
Are there certain climate conditions where asexual reproduction or sexual reproduction might be favored?

How Do Offspring Compare to Parents?

A dog breeder mates a male and female to produce a litter of puppies. Which of these statements is true about the puppies? **Select** the correct answer.

- ○ Some are identical to their mother.
- ○ They share some traits with each parent.
- ○ Some share no traits with their mother.
- ○ They share no traits with either parent.

Structure of a Flower

Label each part of the flower using the terms below.

| filament | style | ovules | stigma |
| anther | pollinator's body | | |

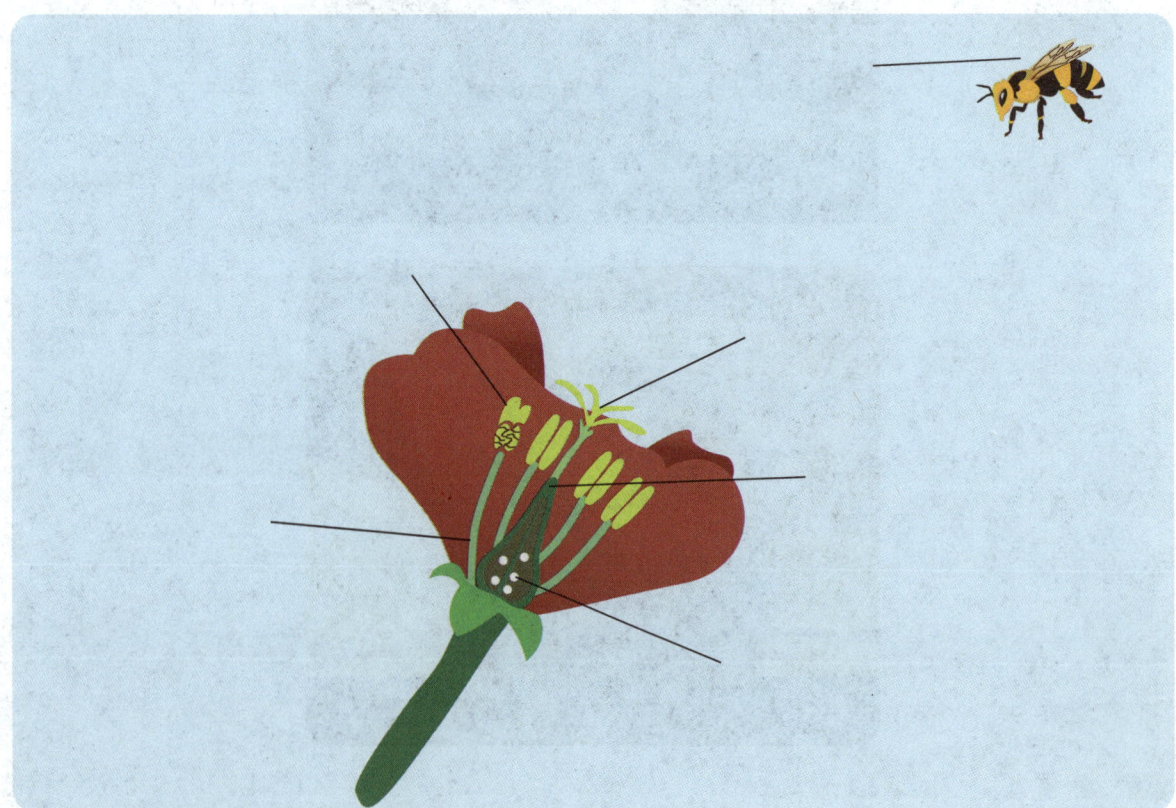

Behavior Classification

Animals have evolved behaviors that help them survive harsh environmental factors as well as behaviors that promote their ability to reproduce to continue the survival of their species. Classify each behavior as either increasing the animal's survival in its environment or increasing its reproductive success by writing the number of the behavior in the correct column of the table.

1. Sea lions remain underwater while listening for sounds of fish moving in their vicinity.

2. Male cardinals chase other male cardinals away from the area where they live.

3. Male monkeys rub urine into their fur.

4. Young giraffes get up and begin walking around 15 minutes after they are born.

5. Male peafowl, known as peacocks, open up their tail feathers to display them.

6. Male polar bears follow scent trails left by female polar bears.

7. Both males and females of many spider species construct intricate webs.

Survival Behaviors	Reproductive Success Behaviors

Concept 3.3: Reproductive Success

How Do Living Things Reproduce?

Activity 4
Analyze

Types of Reproduction

Quick Code
ca6561s

Read the the following text about how living things reproduce. As you read, **highlight** information you can use as evidence to support your initial ideas to answer the Can You Explain? question.

Types of Reproduction

Why do living things **reproduce**? Humans have children for many different reasons, but all organisms reproduce because individuals have limited lifespans. Life continues because organisms are able to reproduce and produce new individuals. Each organism produces **offspring** similar to itself because it passes on some of its genes. Genes are pieces of DNA (deoxyribonucleic acid) found in chromosomes and code for specific traits, such as height or hair color.

Under normal conditions, when organisms reproduce, their offspring have the same number of chromosomes as the parents. For example, a fruit fly only has four chromosomes. When it reproduces, each of the offspring have four chromosomes as well. What can differ is the way that organisms accomplish reproduction.

Asexual reproduction results in offspring that are genetically identical to each other. Only one parent is required. In its simplest form, a single-celled organism divides into two. This process can occur relatively quickly.

In sexual reproduction, two individuals provide half of their genes in a special cell called a gamete to make a complete offspring. This is done when a female gamete, the **egg** or ovum, is fertilized by a male gamete, **sperm**. The egg and sperm each contain half of the chromosomes needed for a complete individual. During **fertilization**, when the egg and sperm combine, the offspring that results has a new combination of genes that neither parent had. This process takes longer and results in a population that has more genetic variety than one reproduced by asexual reproduction.

EXPLORE

Concept 3.3: Reproductive Success | 119

CONCEPT 3.3 | Are there certain climate conditions where asexual reproduction or sexual reproduction might be favored?

Activity 5
Reason

Organism Reproduction

Quick Code
ca6562s

In this investigation, you will use diagrams to model asexual and sexual reproduction and illustrations to explain how the passage of genetic information from one generation to the next differs in these two types of reproduction.

What materials do you need? (per group)

- Two organisms to investigate
- Sticky notes
- Computer
- Printer paper
- Colored pencils
- Scissors
- Studio

SEP Constructing Explanations and Designing Solutions
CCC Structure and Function

Procedure

After selecting your organisms, you will create models (diagrams) that show how the organisms you selected reproduce, with one diagram for each organism.

Your models must do the following:

- Indicate whether reproduction is asexual or sexual.

- Explain why the offspring would have identical or different genetic information from the parent or parents.

CONCEPT 3.3 | Are there certain climate conditions where asexual reproduction or sexual reproduction might be favored?

Reflect

What new information did you learn about sexual reproduction and asexual reproduction?

_____
_____
_____
_____
_____
_____
_____
_____

What did you notice about the offspring produced by each organism?

_____
_____
_____
_____
_____
_____
_____

Why is the process of reproduction different for each organism?

Concept 3.3: Reproductive Success

CONCEPT 3.3 Are there certain climate conditions where asexual reproduction or sexual reproduction might be favored?

Activity 6
Evaluate

Reproduction

Quick Code
ca6563s

One of the characteristics of living things is that they must reproduce. However, the way they accomplish this can be very different. Read the descriptions of the strategies that the following organisms use to reproduce. Then, decide if they use asexual reproduction, which creates offspring that are genetically identical to a single parent, or if they use sexual reproduction, which involves two parents and creates genetically diverse offspring. **Identify** whether each organism listed reproduces sexually or asexually. **Write** "A" for "asexual" or "S" for "sexual" in the right-hand column.

Description	Image	Sexual or Asexual?
Salmon swim great distances from the ocean back to a river or stream where they were born to reproduce. A female lays her eggs in a hole in the stream bed, and a male covers them with sperm.	Salmon	
Alligators breed at night in shallow waters. After using sound to attract one partner, the alligators will court for about a month before mating. After mating, the female lays her eggs along the bank and covers them with vegetation.	Alligator	

SEP	Constructing Explanations and Designing Solutions
CCC	Structure and Function
CCC	Cause and Effect

Description	Image	Sexual or Asexual?
Hydra produce buds from their body wall. These buds grow until they are mature enough to break off and form a new individual that can exist on its own.	Hydra	
The cactus produces blooms that are needed for the plant to reproduce. These blooms attract bats or bees to the plant, which move them from flower to flower, pollinating them. Birds or other animals then eat the fruit to disperse the seeds to a new location for planting. Fruit that has dried on the cactus plant may use the wind to disperse seeds.	Cactus Bloom	
Bacteria reproduce through a process called binary fission. Within a single bacterium, one DNA molecule will replicate and both copies attach to the cell membrane. As the cell grows, the DNA molecules grow farther apart. When the cell is about twice its original size, it splits into two cells and each of these start the reproduction process over again.	Bacteria	
Starfish can reproduce through a process called fragmentation. In the center of a starfish, there is a central disc. When part of the central disc and a limb breaks off the starfish, it will regenerate into a new starfish.	Starfish	

How Do Animal Behaviors and Plant Structures Affect the Probability of Successful Reproduction?

Activity 7
Analyze

Animal Behavior

Quick Code
ca6564s

Read the text and **highlight** evidence that helps answer the question: How do animal behaviors and plant structures affect the probability of successful reproduction?

_____
_____
_____
_____
_____
_____

Animal Behavior

On a cellular level, **sexual reproduction** is the same in both plants and animals. However, the ways that animals get their **gametes** together are very different.

In most animals, males deposit their gametes (sperm) on or near the female gamete (an egg, or ovum) for fertilization. Typically, land animals mate internally, and aquatic animals mate externally. In external fertilization, the females lay eggs, and the males release sperm onto the eggs. Despite the differences in the mating processes, almost all animals try to attract mates using specific behaviors, physical traits, or a combination of the two. In some animals, both sexes are found in one individual. Most of the time, these animals also must find a mate, and one individual acts as the male and the other the female, or they act as both for the other individual. Very few of these animals are able to fertilize themselves.

For animals to reproduce, each of the two individuals have to convince the other that they are a worthy mate. This means making sure that they are the same species and that they are healthy and will produce well-adapted offspring. Sometimes, the female puts in more effort, but usually the male animal is the one who works hardest. The different physical traits or behaviors can signal to potential mates how intelligent, strong, or desirable an individual is. For example, the male peacock has a large flashy tail. This tail slows him down when he flies and is not helpful in escaping predators. If the peacock is still alive when mating season comes, his tail indicates to peahens that he has the strength and intelligence to survive. If an individual can perform an elaborate mating dance, it shows that he can perform intense physical feats. Many mating rituals are guided by **instinct**—the animals are not taught the **behavior**, but it comes naturally to them. What are some of the different behaviors and traits animals have for attracting mates?

Activity 8
Analyze

Attracting a Mate

Quick Code
ca6565s

Read the following text about some animals' curious behaviors that help in their quest for a mate.

Attracting a Mate

Organisms that reproduce sexually often need to find a mate. Many of these organisms have adaptations that improve their chance of success. Many organisms have structures that improve their chance of successful reproduction. Some organisms have behavioral adaptations that help them attract a mate.

Mourning Cuttlefish

Cuttlefish are cephalopods, which are invertebrates that live in the ocean. Cuttlefish have millions of chromatophores in their skin. Chromatophores are cells of pigment that are attached to tiny muscles. Cuttlefish control chromatophores to change the color and pattern of their skin. This helps them hunt, hide, and find mates. When scientists studied how mourning cuttlefish (*Sepia plangon*) use color to attract mates, they made an amazing discovery. In very small groups, some male mourning

Mourning Cuttlefish

cuttlefish will display both female and male coloring to find mates. A male cuttlefish will position itself between a female and another male. The half of the middle cuttlefish's body that faces the female will have male coloring. The half that faces the male cuttlefish will have female coloring. Because the second male sees female coloring, it doesn't recognize that another male is competing for the female. As a result, the middle male has more time to attract the female. This increases the chances that the middle male will mate with the female.

Weaver Birds

Weaver birds are small seed-eating birds. Most species of weaver birds live in Africa. Like many birds, weaver birds build nests. But for weaver birds, nests play a key role in attracting a mate. Male weaver birds use grass and other materials to build a complex hanging nest. For many weaver bird species, the entrance to these nests is often at the bottom. The male weaver bird hangs upside down from the entrance. He flaps his wings and calls to attract female birds to his nest. The female bird inspects the nest. Female birds are less likely to mate with younger males. Young male weaver birds are still learning how to build nests. Their nests are not as strong or as well positioned as those of older birds. Therefore, mature male weaver birds tend to be more successful at attracting a mate.

Weaver Birds

Attracting a Mate *cont'd*

Elk

Elk are social herbivores that spend most of the year split into two groups. One group includes mainly females and immature males. The other group is made up of mature male, or bull, elk. During the fall mating season, the bulls will join the female group for a short time. A bull elk will gather a group of females, or harem. To reproduce successfully, a bull elk must keep other males away from his harem. Antlers make this possible. Bulls grow a new set of antlers each year, starting in spring. By the time mating season starts, their antlers are solid bone. A bull elk uses his antlers to fight other males and predators. Elk that are strong and have large antlers are usually more successful at fighting off other males. As a result, they usually attract more mates.

Elk

Answer the questions below based on the passage.

In the passage, which adaptations for finding a mate are structural? Which are behavioral?

What is another example of a structure or behavior that helps an organism attract a mate?

_____
_____
_____
_____
_____
_____
_____

In many species, the male is much larger than the female. How might this improve reproductive success?

_____
_____
_____
_____
_____
_____
_____
_____

Activity 9
Observe

Flower Power

Quick Code
ca6566s

Go online to discover how a flower's color, scent, and shape attract pollinators like bees, bats, and butterflies.

What did you learn about how flowers attract pollinators?

_____
_____
_____
_____
_____

SEP Constructing Explanations and Designing Solutions

CCC Structure and Function

Concept 3.3: Reproductive Success

Activity 10
Analyze

Quick Code
ca6567s

Pollination: To Bee or Not to Bee

Read the following passage. **Look** for details about flower characteristics and their purpose. Then, **complete** the activity that follows.

Pollination: To Bee or Not to Bee

A flower's appearance and structure greatly affect how it is pollinated. Pollination begins when pollen grains from the male part of the flower land on the female part. In most cases, the wind or animal pollinators help the process.

What's the Buzz?

Insects like, and even depend on, flowers. Butterflies, moths, and particularly bees get most of their nutrition from the flowers they visit. To attract these guests, some plants produce nectar, a sweet-smelling, sweet-tasting substance. As the insect drinks the nectar, it might pick up bits of pollen dust on its fuzzy body. When it stops at another plant, it leaves some of the pollen behind, and fertilization begins.

Sooo Sweet

To protect their valuable content, flowers that produce nectar have evolved into many shapes. Often, the petals of nectar-producing flowers form funnels, bells, and tubes—with the goodies hidden deep inside. Bees, butterflies, and moths eat the nectar; some use their long tube-shaped mouths and long tongues.

Tough It Out

Mammals get into the pollinating act, too. Some bats pollinate the giant saguaro and other cactus flowers. In tropical areas, many plants depend on the long nose and tongue of the bat to pollinate the flowers that produce bananas, avocados, mangoes, and guavas. Bats have poor eyesight, yet they do have strong claws to cling to the plants. This is why the plant species pollinated by bats must be tough and large.

Masquerade Ball

Some orchids actually look and smell like the female mates of certain wasps and flies. They can even carry the fragrance female insects give off when they are ready to mate. Eager males follow the scent and try their hardest to complete the mating dance. On their next visit to a flower, they shake off the orchid pollen stuck to their bodies!

Ah-Choo!

In the summer, when your eyes and nose start itching, it could be from the pollen grains floating through the air. The flowers of many wind-pollinated plants, such as grasses, are often pale and small, and have no fragrance. These plants depend on breezes, not animals, to carry on their family tradition.

Seeing Red

Brightly colored flowers appeal to animals that are awake and hungry during the daytime. Bees buzz most around flowers that are blue and yellow, while hummingbirds take a direct route to red waxy flowers. Scientists believe birds can see the vivid color red, and insects see a paler version that is not as rosy. Pale petals stand out against the darkness, so flowers pollinated by moths and bats are often very light colored or white.

Pollination: To Bee or Not to Bee *cont'd*

Trees in the Breeze

Have you ever seen a helicopter land right at your feet? Chances are you have, especially if you have been around pine trees. Most pine trees carry both male and female cones. Pollen dust is released by the male cone and is carried by the wind. Some of it gets stuck on the female cones, causing the cone to grow to four times its size. Now packed with seeds, the mature cone opens, releasing seeds that have a winglike scale attached. As the seeds fall, they spin through the air like helicopters, eventually landing in an area where a new pine tree can grow.

Create a six word story about what you have read.

_____
_____
_____
_____
_____
_____
_____
_____
_____
_____
_____
_____
_____

| CCC | Cause and Effect |
| CCC | Structure and Function |

CONCEPT 3.3 | Are there certain climate conditions where asexual reproduction or sexual reproduction might be favored?

Activity 11
Evaluate

Carried Away and Climate Impacts

Carried Away

Quick Code
ca6568s

Use your understanding of plant features to determine if the following plants are pollinated by animals or wind. **Write** either "wind" or "animals" in the right column.

Plants	Pollinated by Wind or Animals?
(corn)	
(pink daisy)	
(wheat)	

138

Plants	Pollinated by Wind or Animals?

Concept 3.3: Reproductive Success | 139

CONCEPT 3.3 | Are there certain climate conditions where asexual reproduction or sexual reproduction might be favored?

Climate Impacts

Use the bar graph to help you answer the questions.

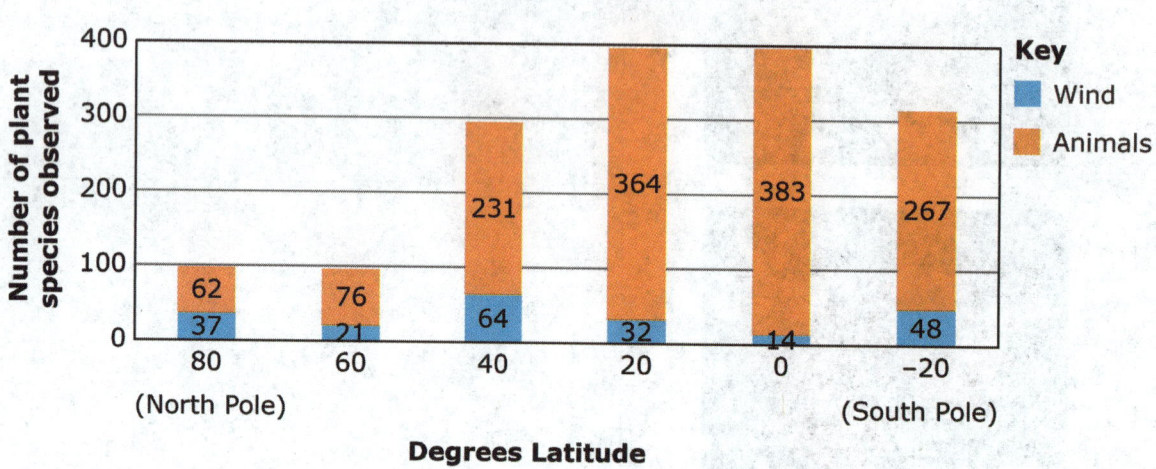

Consider the different climate conditions found in the latitudes listed in the bar graph to provide a possible reason for the distribution of wind and animal pollination data in the graph.

___
___
___
___
___

Is one climate more favorable than another for pollination?

CONCEPT 3.3 — Are there certain climate conditions where asexual reproduction or sexual reproduction might be favored?

Activity 12
Solve Problems

Quick Code
ca6570s

Build a Better Pollinator

Insects are expert pollinators for plants, but often they are regarded as pests by humans. Imagine that you have been challenged to design a perfect pollinator. Based on what you have learned so far, **answer** the questions below.

Using what you have learned about the relationship between plants and pollinators, as well as your knowledge of the structure of a flower, **write** a description of an ideal pollinator. Include information on the size of your pollinator and the features that would make it successful.

Pollinator

_____
_____
_____
_____
_____

SEP — Constructing Explanations and Designing Solutions

CONCEPT 3.3 | Are there certain climate conditions where asexual reproduction or sexual reproduction might be favored?

Humans have devised artificial methods of pollination to supplement or replace insect pollination. **Conduct** online research on artificial pollination methods. **Summarize** three of the pros of these technologies as well as three cons to these technologies. Then, **explain** which technology you believe to be the most useful, and why.

Select another pollinator and do some research on how that pollinator is well adapted to the types of flowers it pollinates. **Write** a description of how the pollinator's design matches the flower's design.

CONCEPT 3.3 | Are there certain climate conditions where asexual reproduction or sexual reproduction might be favored?

Activity 13
Record Evidence

Nests and Hatchlings

As you worked through this lesson, you investigated and gathered evidence about reproductive success. Now, take another look at the Nests and Hatchlings video, which you first saw in Engage.

Let's Investigate Nests and Hatchlings

How has your understanding of the Nests and Hatchlings video changed?

Read the Can You Explain? question from the beginning of this lesson.

Quick Code
ca6571s

 Can You Explain?

Are there certain climate conditions where asexual reproduction or sexual reproduction might be favored?

Use your new understanding of the Nests and Hatchlings video to write a scientific explanation answering a question. Recall that a scientific explanation contains three elements: a scientific claim, evidence to support the claim, and reasoning that connects the evidence to the claim. Be sure to include evidence that describes qualitative or quantitative data that can be used to support your claim.

SEP Constructing Explanations and Designing Solutions

146 |

Write your scientific explanation in the space provided.

Use these sentence starters to help you construct your scientific explanation.

My claim is . . .

CONCEPT 3.3

Are there certain climate conditions where asexual reproduction or sexual reproduction might be favored?

The evidence I found . . .

STEM in Action

Activity 14
Analyze

Genetic Engineering and Agriculture

Quick Code
ca6572s

Read the text and **watch** the videos. As you read, **highlight** details about how scientists modify plant genes, and what problems genetic engineering might be used to solve. Then, **complete** the activities that follow.

Genetic Engineering and Agriculture

Genetic engineering is the process of manipulating an organism's genetic material in an effort to produce desired traits. The genetic engineering of plants is an extremely important field in the agricultural industry. Most of the food we eat comes from organisms that humans have modified in some way. To modify plants, genetic engineers identify the genes for desirable traits and insert them into new plants. Examples of desirable traits in a plant might be fruit that is sweeter, larger in size, or has fewer seeds. Other desirable traits include resistance to drought and pests.

Video — Creating Transgenic Plants

Plants that have been genetically engineered are called transgenic plants. The process of asexual reproduction plays an important role in genetic engineering. The video explains more.

Researchers recently have begun to consider the role of asexual reproduction in the development of genetically engineered plants in a new way, and they are working to learn enough about the genetics of asexual reproduction to apply it to plants that reproduce sexually. Turning sexually reproducing plants into asexual reproducers could have profound implications for agriculture.

Currently, farmers throughout the world spend billions of dollars each year to buy seeds for crops. They can't grow these seeds themselves because the process of sexual reproduction erases many desired traits such as resistance to drought or pests. So every year, farmers must purchase new supplies of seeds. Seeds that are produced asexually require less labor to produce. These seeds could save farmers a lot of money.

Transgenic Plants: Activism and Regulation

Genetic Engineering and Agriculture *cont'd*

For this reason and others, transgenic plants have been hailed by some as a critical tool for solving the world's agricultural problems. Genetically engineered plants already allow humans to produce food more efficiently. Farmers can generate higher crop yield using less space. New discoveries in genetic engineering, like seeds produced asexually, might solve even bigger agricultural problems—like world hunger.

Genetically engineering plants remains a heavily debated topic, however. Potentially unknown health and environmental risks of genetically engineered crops concern some biologists, ecologists, and activists.

How do you feel about genetic engineering? On what information, statistics, and beliefs do you base your opinion?

CONCEPT 3.3 Are there certain climate conditions where asexual reproduction or sexual reproduction might be favored?

Which Images?

These four images are artistic creations. They are meant to illustrate the concept of genetically engineered foods. Study the images carefully. Then, answer the questions that follow.

Apple-Orange Hybrid (A)

Tomato (C)

Protective Gear (B)

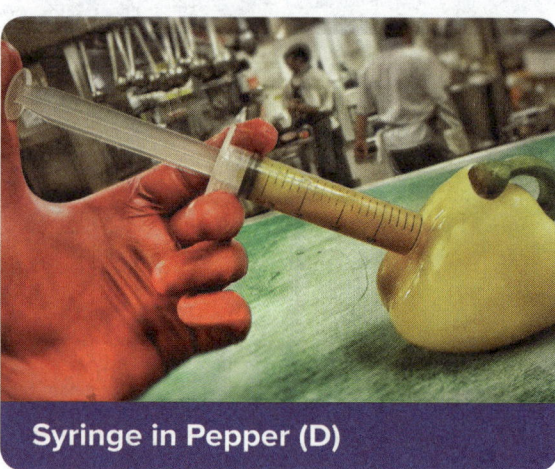
Syringe in Pepper (D)

Choose one image from each pair: Images A or B, and Images C or D. What does each image show? How does each one make you feel?

_____

_____

_____

_____

Looking Critically

Analyze and compare the images you selected more carefully. Why do they make you feel what you feel? Cite specific details as you analyze and compare.

_____
_____
_____
_____
_____
_____
_____

What Is the Message?

Use your analysis to infer the artist's purpose in creating each image. What do you think the artist wants the viewer to think and feel about genetically engineered foods? What evidence from the images you selected supports your ideas?

_____
_____
_____
_____
_____
_____

CONCEPT 3.3

Are there certain climate conditions where asexual reproduction or sexual reproduction might be favored?

Activity 15
Concept Review

Review: Reproductive Success

Quick Code
ca6574s

Now that you have completed the objectives for this concept, review the core ideas you have learned. Record some of the core ideas below.

Core Ideas

Talk with a Group

Now, think about the annual average temperature map you saw in Get Started. Discuss how what you've learned about reproductive success can help you understand the annual average temperature map.

Concept 3.3: Reproductive Success | 157

CONCEPT 3.4

Heredity

Student Objectives

By the end of this lesson:

- ☐ I can develop models that describe and predict patterns in the inheritance of dominant and recessive traits.
- ☐ I can construct explanations as to how the laws of inheritance account for stability and change in phenotype from generation to generation.
- ☐ I can develop and use a model to predict the probable genotypes of offspring based on the genotypes of the parents.
- ☐ I can argue from evidence that many traits show patterns of inheritance other than Mendelian inheritance.
- ☐ I can represent data as ratios to show constant proportions within large numbers of offspring.

Key Vocabulary

allele, dominant trait, F1 generation, gene, generation, genotype, heredity, heterozygous, homozygous, inherit, offspring, parent, phenotype, Punnett square, recessive

Quick Code
ca6576s

Concept 3.4: Heredity

Activity 1
Can You Explain?

How could a greater diversity of traits help a population adapt to a wider range of climates?

_____

_____

_____

_____

_____

_____

_____

_____

_____

Quick Code
ca6577s

CONCEPT 3.4 | How could a greater diversity of traits help a population adapt to a wider range of climates?

Activity 2
Ask Questions

Maintaining Potato Biodiversity

Quick Code
ca6578s

Watch the video and **answer** the questions that follow.

Let's Investigate Maintaining Potato Biodiversity

SEP Obtaining, Evaluating, and Communicating Information

162 | Discovery Education

Why was the International Potato Center formed?

_____
_____
_____
_____

How do you think reproduction played a role in creating native and improved potatoes?

_____
_____
_____

Make a list of questions that would have "genetic variation" as the answer.

_____
_____
_____
_____
_____

What questions do you have about biodiversity?

_____
_____
_____

Concept 3.4: Heredity | 163

CONCEPT 3.4 | How could a greater diversity of traits help a population adapt to a wider range of climates?

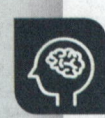

Activity 3
Evaluate

Quick Code
ca6579s

What Do You Already Know About Heredity?

Black and White Rabbits

Read the scenario, and then **answer** the question.

A pure black rabbit is going to be crossed with a pure white rabbit. The breeder tells you that black fur is dominant and white fur is recessive. What will the rabbits resulting from this cross look like?

Black Rabbits

Answer the following question.

If black fur is dominant in rabbits, is it ever possible for two black rabbits to produce white offspring? Explain your answer.

CONCEPT 3.4 How could a greater diversity of traits help a population adapt to a wider range of climates?

Punnett Square

The Punnett square below shows one allele from each parent. **Fill in the blanks** to complete the Punnett square.

	B	
	Bb	
b		Bb

The Next Generation

Look at your results from the Punnett square in the previous item. Suppose that B is the dominant trait and b is the recessive trait. What are the chances that the offspring of these two parents will display the recessive trait?

What Is Heredity, and How Were the Principles of Heredity Discovered?

Activity 4
Observe

Quick Code
ca6583s

Family Tree

Study the family tree below, and then **answer** the question that follows.

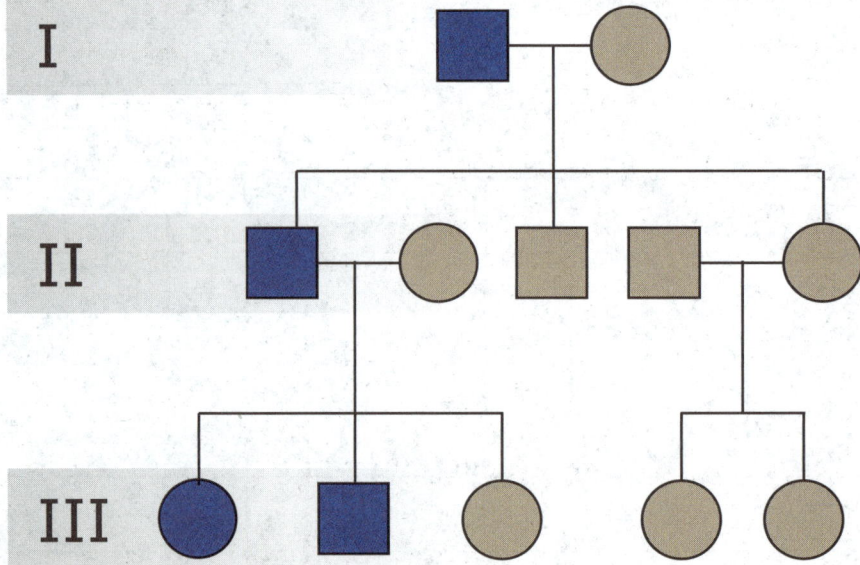

The family tree shows that certain traits are shared, and some are not. What would cause siblings to inherit different traits from the same parents?

_____

_____

_____

_____

CCC Cause and Effect

Activity 5
Analyze

Principles of Heredity

Quick Code
ca6584s

Read the text. **Highlight** information you can use as evidence to support your initial ideas on how to answer the Can You Explain? question or other questions you may have about this topic.

Principles of Heredity

In the middle of the 19th century, a European monk named Gregor Mendel became interested by the idea that traits can be passed from parents to their <mark>offspring</mark>. He was aware that selective breeding was done to produce plants and animals with desirable traits. At that time, people thought that traits displayed by offspring represented a blend of the parents' traits. Mendel designed and ran a large number of experiments using pea plants to investigate this idea. His goal was to establish patterns of inheritance. Then, he could use the patterns to explain how traits were passed from one <mark>generation</mark> to the next. This passage of traits through generations is called <mark>heredity</mark>.

Gregor Mendel

Principles of Heredity *cont'd*

Mendel ran controlled experiments with thousands of pea plants. He transferred pollen from one plant to the flowers of another plant so that seeds would be produced. These experiments were called "crosses." The plants that were crossed in this way are called the parental generation. The seeds produced by the cross were planted to produce offspring known as the **F1 generation**. When the F1 plants matured, they could be crossed with one another to produce a third generation. The third generation is called the F2 generation.

Mendel found that when he crossed pea plants that produced round seeds with pea plants that produced wrinkled seeds, all individuals in the F1 generation had round seeds. When he crossed two individuals from the F1 generation, the F2 generation had individuals that produced round seeds and individuals that produced wrinkled seeds. Moreover, there was a 3:1 ratio in the number of F2 individuals with round seeds compared to the number with wrinkled seeds.

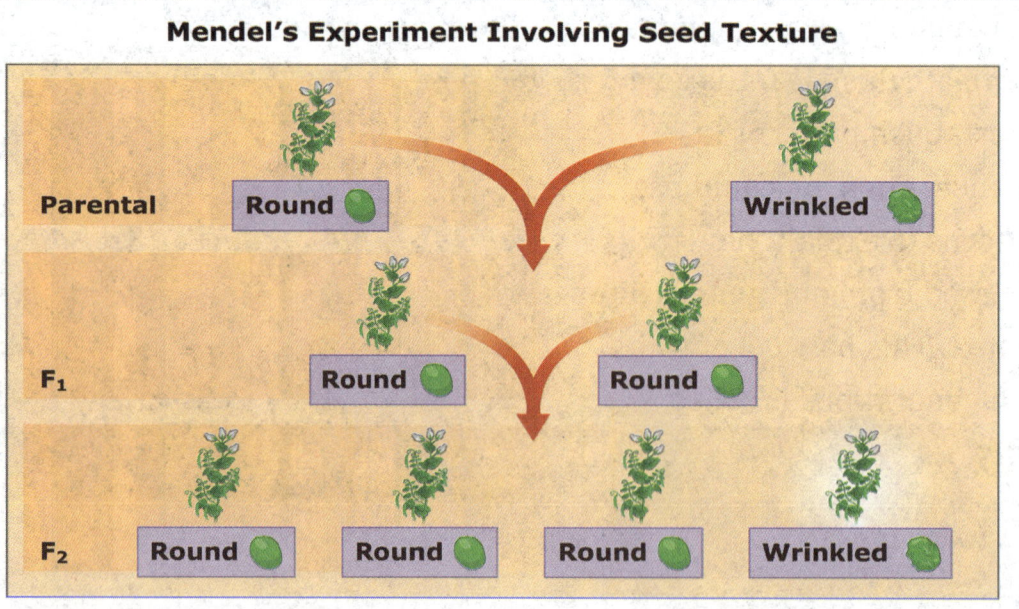

Mendel carried out many experiments in a systematic fashion. He kept careful track of the numbers of offspring produced in each cross and their traits. Because of his attention to detail and careful recordkeeping, Mendel was able to show that specific patterns of inheritance were reproducible. That means that the same pattern of inheritance is observed repeatedly from a particular type of cross. From his results, Mendel was also able to conclude that offspring **inherit** two factors for each trait, one from each **parent**. If an individual inherits two different factors for a particular trait, only one factor will be expressed. However, the factor not expressed is still present in that individual. That factor can be passed to the next generation, where it may be expressed.

These ideas are fundamental to understanding the mechanism of heredity. Scientists have learned more about heredity since Mendel's time. For example, we now know that the genes that produce biological traits are carried with other genes on structures called chromosomes. Each individual cell contains an even number of paired chromosomes—humans have 23 pairs. In sexual reproduction, an individual inherits one copy of a chromosome, and therefore one copy of each **gene**, from each parent. These genes may be different or the same. For example, a gene that codes for hair color may have different forms. These different forms of a gene coding for a particular trait are called alleles.

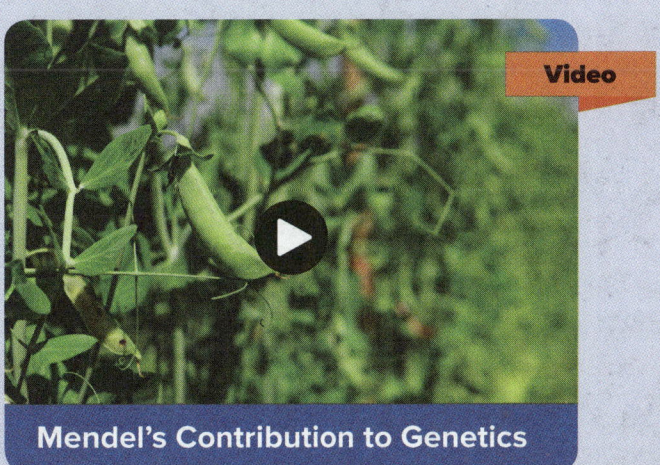
Video
Mendel's Contribution to Genetics

Think about what you have read and watched. Using the graphic organizer, **summarize** the information based on the text and video.

Topic: _____

Activity 6
Evaluate

Explaining Trait Patterns

Read the scenario, and then **answer** the question.

Quick Code
ca6585s

A woman buys a packet of seeds for red flowering plants. In the spring, she plants the seeds in her garden and enjoys beautiful red flowers all summer. The flowers are self-pollinating, and many of the flowers produce seeds. The woman collects these seeds and plants them in her garden the following spring. She is surprised when about one-fourth of these plants produce white flowers while the other three-fourths produce red flowers. If the seeds were all produced by self-pollination of the parental generation, what would be the reason for the woman's observations?

Concept 3.4: Heredity

How Do Scientists Make Predictions About Genetic Inheritance and Expression of Traits?

Activity 7
Analyze

Quick Code ca6586s

Representing Alleles

Read the following text about genetic inheritance and the expression of traits. As you read the text and watch the video, gather evidence to support or refute the following statements about genetic inheritance. **Write** the evidence in the table provided.

Statement 1: I get my long legs from my father.
Statement 2: I get my curly hair from my grandmother.

Representing Alleles

Alleles and Punnett Squares

Our genes determine many of our traits. Each person inherits two copies (or alleles) of each **gene**—one from each **parent**. You can create a model called a **Punnett square** to represent the two alleles for a gene in a mother and the two alleles for the same gene in a father. Letters of the alphabet used to represent these alleles are written on the top and the side of the Punnett square. A capital letter represents one **allele**, while a lowercase letter represents a different allele for a

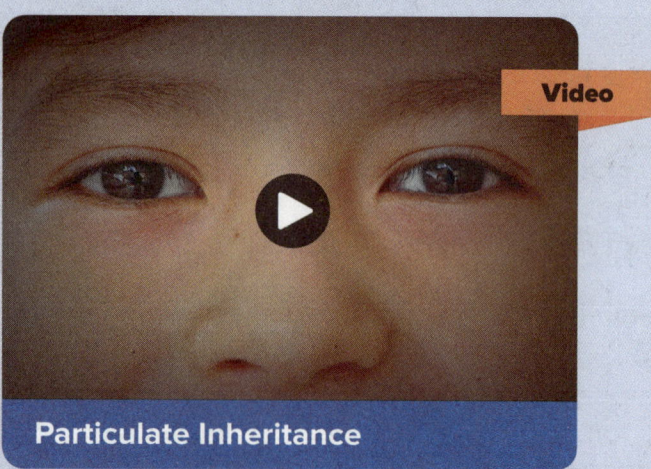

Particulate Inheritance (Video)

gene. The body of the Punnett square is filled in with the allele combinations that can occur in the children of the two parents. The Punnett square is a simple model that can be used to predict all the possible ways that alleles can be passed from parents to **offspring**.

Dominant and Recessive Genes

When offspring **inherit** two alleles that are identical, they are said to be **homozygous** for that trait. For example, a homozygous combination could be represented as "AA" or "aa," where "A" is one allele for a gene and "a" is another allele for the same gene. If offspring inherit two different alleles, they are said to be **heterozygous** for that trait. A heterozygous combination for the previous example would be represented as "Aa."

Mendel found that heterozygotes do not show a blend of the two traits specified by their different alleles. Instead, they show only one trait specified by one allele and none of the trait specified by the other allele. The allele corresponding to the displayed trait is said to be dominant. The allele corresponding to the trait that is not displayed is said to be **recessive**. Dominant alleles are represented with capital letters, and recessive alleles are represented with lowercase letters. For example, the allele for purple flower color in pea plants is dominant and represented with the letter "P." The allele for white flower color is recessive and represented with the letter "p." A pea plant heterozygous for this trait would be represented as "Pp" and would display purple flowers. A pea plant homozygous for the dominant allele would be represented as "PP" and would also have purple flowers. The only allele combination that would produce white flowers would be the homozygous "pp" combination because the white flower trait is recessive.

Write the evidence you gathered from the text and video in the table below.

Supporting Evidence	Refuting Evidence

 **Activity 8**
Observe

Punnett Square

The image shows a Punnett square that models the cross of two heterozygous parents. **Study** the image, and then **answer** the question that follows.

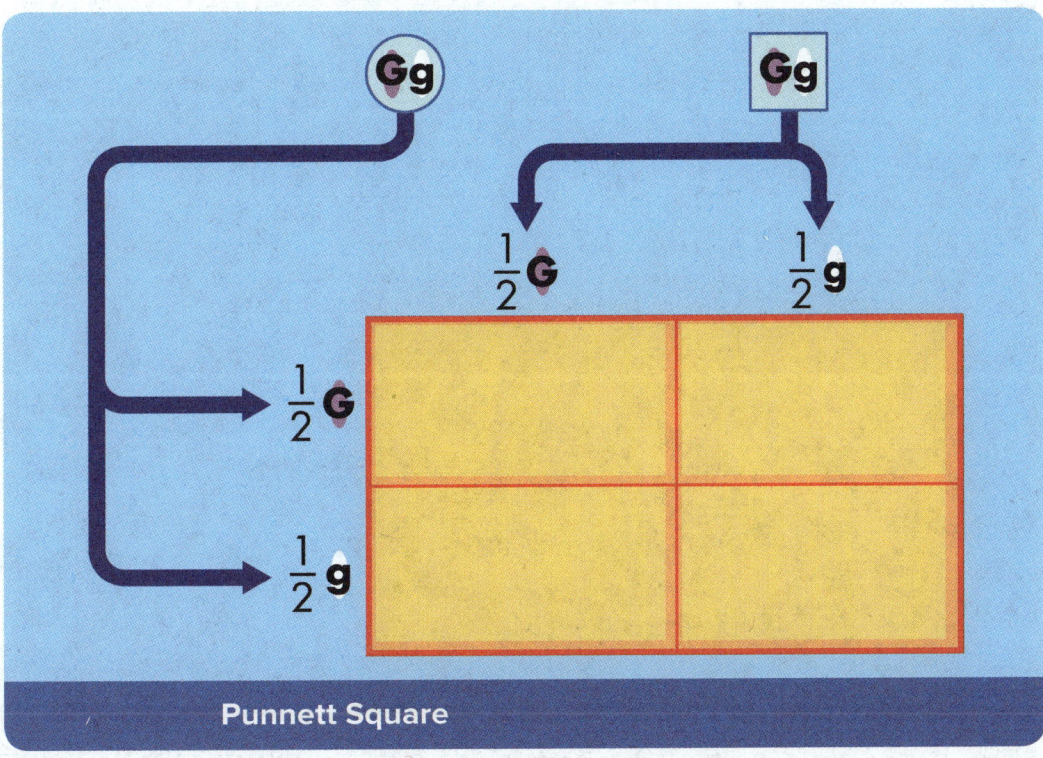

Punnett Square

CONCEPT 3.4

How could a greater diversity of traits help a population adapt to a wider range of climates?

Determine the genotypes of the offspring.

	G	G
g	Gg	Gg
g	Gg	Gg

	G	g
g	Gg	gg
g	Gg	gg

Activity 9
Analyze

Quick Code
ca6588s

Genotypes and Phenotypes

Read the text. As you read, **highlight** the differences between genotype and phenotype.

Genotypes and Phenotypes

Some combinations of alleles differ yet produce the same trait in an individual. This is because there are dominant and recessive forms of alleles. To help make these cases clear, scientists distinguish between **genotype** and **phenotype**. Genotype is a description of the alleles present in an individual. Phenotype is the outward expression of a trait in an individual. The genotypes "AA" and "Aa" differ, yet both produce the same phenotype. This is because the dominant "A" allele is expressed in both cases.

Genotypes and Phenotypes

Genes	Alleles	Genotype	Phenotype
Shape	R	RR	Round
		Rr	Round
	r	rr	Wrinkled
Height	T	TT	Tall
		Tt	Tall
	t	tt	Short
Color	P	PP	Purple
		Pp	Purple
	p	pp	White

Recessive phenotype: Results from genotype with two recessive alleles

CCC Cause and Effect

Concept 3.4: Heredity   179

Activity 10
Analyze

Quick Code
ca6621s

Genes That Influence More Than One Trait

Before you read the passage, think about the following questions:

- How can the great variety of eye colors in humans be explained?
- Describe how the human gene *ABCC11* affects both underarm odor and earwax.

Highlight words in these two questions that you should look for while reading. Then, **read** the text and **highlight** details that will help you answer the questions.

Genes That Influence More Than One Trait

Not all inheritance occurs as simple Mendelian inheritance.

Most traits are determined by more than one gene, and many traits have more than two alleles. For example, the great variety of eye colors can only be explained by the action of several sets of genes and alleles.

Another example is found on the human chromosome number 16.

The *ABCC11* gene marked in red on the following chromosome drawing is another example of this type of gene.

The protein-creating instructions embedded by this gene govern many different bodily processes. The proteins regulate the way that other molecules travel around the

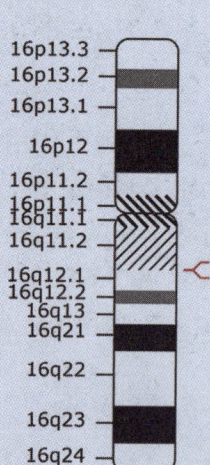

body. In this case, the proteins control which molecules are allowed to enter or leave cells. One version of this gene causes people to have less underarm body odor because those molecules are not allowed to leave the cells! Another version will cause a person to have a different kind of earwax. It is all because of the various alleles and the body processes that they regulate.

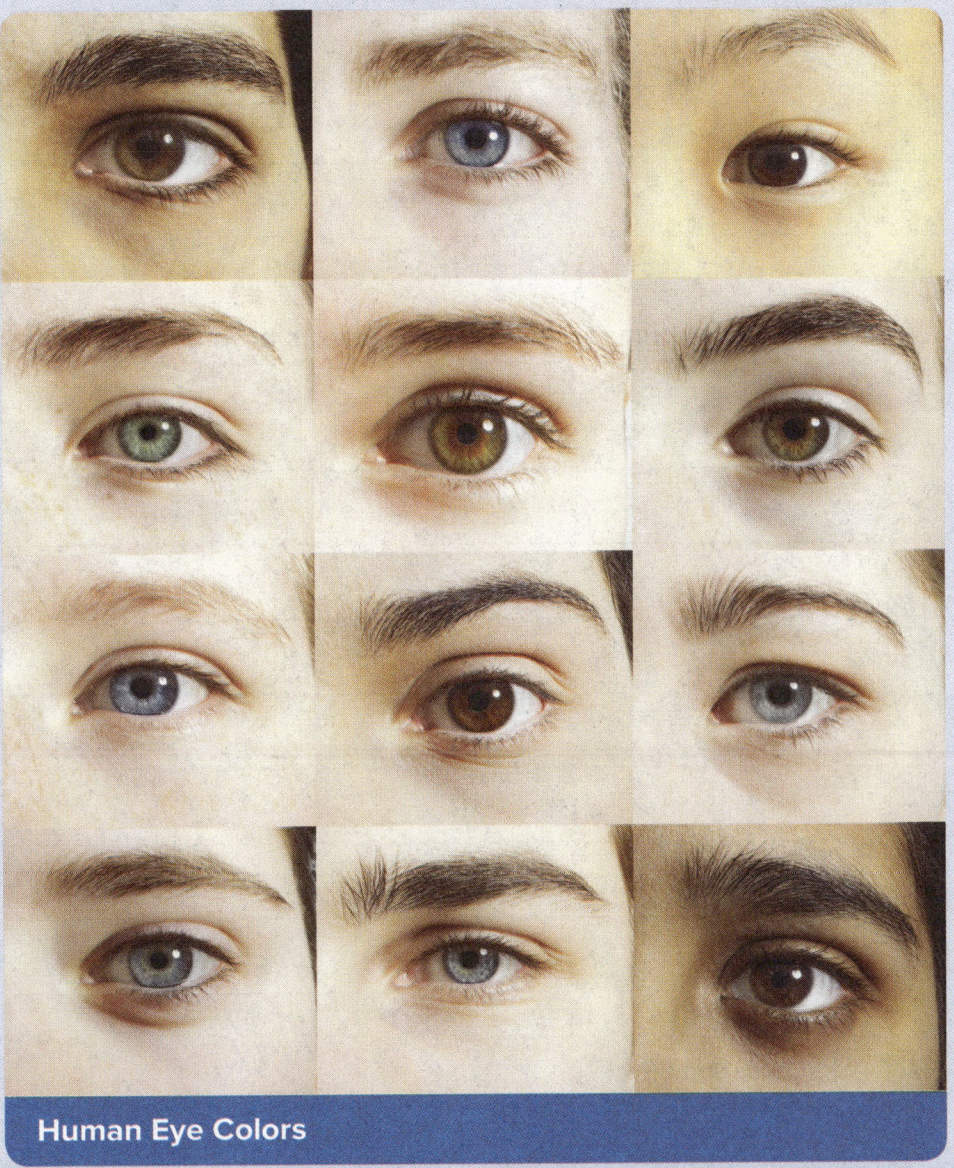

Human Eye Colors

Now, **respond** to the questions. **Share** your answers with the class.

How can the great variety of eye colors in humans be explained?

_____
_____
_____
_____
_____
_____

Describe how the human gene *ABCC11* affects both underarm odor and earwax.

_____
_____
_____
_____
_____
_____
_____

CCC **Cause and Effect**

 **Activity 11**
Observe

Breeding Pea Plants

Quick Code
ca6589s

Go online to explore how to use Punnett squares to analyze and interpret possible outcomes from genetic crosses for pea plants. Then, **answer** the questions that follow.

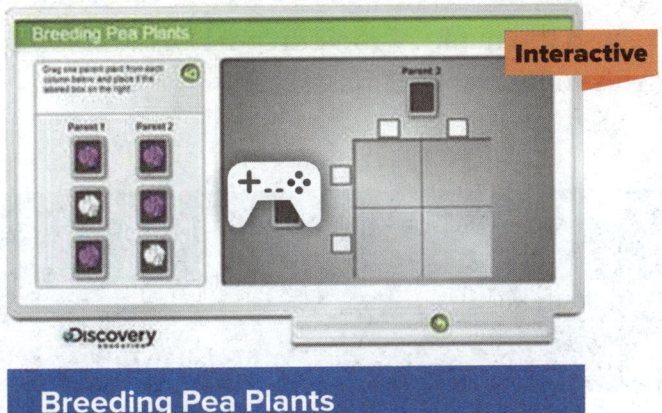

Breeding Pea Plants

When can two plants with purple flowers produce offspring with white flowers?

_____

_____

_____

Can two plants with white flowers produce offspring with purple flowers? Explain.

_____

_____

_____

SEP Using Mathematics and Computational Thinking

SEP Analyzing and Interpreting Data

CONCEPT 3.4
How could a greater diversity of traits help a population adapt to a wider range of climates?

When a Pp-genotype plant is bred with a pp-genotype plant, what percentage of the offspring is likely to have white flowers?

_____

_____

_____

In the Data Chart, **record** the number and percentage of offspring with different combinations of alleles for each set of parents.

Parents		Offspring					
		PP (Purple)		Pp (Purple)		pp (White)	
1	2	Number	%	Number	%	Number	%
Pp	PP	2	50	2	50		
Pp	Pp						
PP	PP						
PP	pp						
pp	pp						

Activity 12
Evaluate

Analyzing Hair Length and Analyzing Hair Color

Quick Code ca6590s

Analyzing Hair Length

Two genetic crosses have been proposed between male and female mice with varying hair length. These traits are described in the table below. **Draw** a Punnett square for a cross between (1) male BB and female bb parents and (2) male Bb and female Bb parents.

Allele	Dominant or Recessive?	Trait Description
B	dominant	short hair
b	recessive	long hair

SEP Developing and Using Models

CONCEPT 3.4 — How could a greater diversity of traits help a population adapt to a wider range of climates?

Using the Punnett squares you created, **answer** the following questions.

Identify the phenotypes and genotypes produced by the cross between a male BB and a female bb.

_____

_____

_____

_____

_____

_____

Identify the phenotypes and genotypes produced by the cross between a male Bb and a female Bb.

_____

_____

_____

_____

_____

_____

_____

_____

Analyzing Hair Color

Two genetic crosses have been proposed between male and female mice with varying hair color. These traits are described in the table. **Draw** a Punnett square that models the cross between a male parent gg and female parent Gg. Then, **draw** another Punnett square that models the cross between a male parent Gg and female GG parent.

Allele	Dominant or Recessive?	Trait Description
G	dominant	gray
g	recessive	black

Using the Punnett squares you created, **answer** the following questions.

Identify the phenotypes and genotypes produced by the cross between a male gg and a female Gg.

_____

_____

_____

_____

Identify the phenotypes and genotypes produced by the cross between a male Gg and a female GG.

_____

_____

_____

_____

Concept 3.4: Heredity

CONCEPT 3.4 — How could a greater diversity of traits help a population adapt to a wider range of climates?

Activity 13
Record Evidence

Maintaining Potato Biodiversity

Quick Code
ca6591s

As you worked through this lesson, you investigated and gathered evidence about heredity. Now, take another look at the Maintaining Potato Biodiversity video, which you first saw in Engage.

Let's Investigate Maintaining Potato Biodiversity

How has your understanding of Maintaining Potato Biodiversity changed?

Read the Can You Explain? question from the beginning of this lesson.

> **Can You Explain?**
>
> How could a greater diversity of traits help a population adapt to a wider range of climates?

Use your new understanding of Maintaining Potato Biodiversity to write a scientific explanation answering a question.

SEP Constructing Explanations and Designing Solutions

SEP Engaging in Argument from Evidence

Write your scientific explanation in the space provided.

Engaging in Argument Choose a scientific explanation from one concept in the unit. Compare and critique the responses of your group members. Did they use the same evidence? Did they interpret the facts the same way you did? How could your response have been stronger?

 in Action

Activity 14
Analyze

Careers in Genetics

Quick Code
ca6592s

Read the text and **view** the video. As you read, **highlight or underline** details about karyotypes and how they are used to diagnose conditions.

Careers in Genetics

Scientists unraveled the blueprint of life by analyzing DNA from humans and other species. They discovered that humans and animals share many of the same genes. For example, a comparison of human and chimpanzee DNA shows a 98 percent match. Mice share 92 percent of the genes found in humans.

Scientists also have used DNA studies to determine whether certain diseases have a genetic basis. In cases where a disease has a genetic basis, scientists find that specific genes are affected. They identify these genes by comparing DNA from people affected by the disease and people not affected.

A human geneticist or medical geneticist is a scientist who analyzes human DNA to diagnose medical conditions. These scientists also help determine the probability that an **offspring** will have a certain medical condition and work to find cures for such conditions.

They often carry out their analyses using blood samples from patients. Watch the video to learn more about the job of a medical geneticist named Wendy Rubinstein.

Geneticist

Most of the samples Wendy collects in her clinic are blood samples, but geneticists also may analyze bone marrow, amniotic fluid, cheek cells, or saliva. A sample may contain enough DNA to study easily. However, some samples contain amounts of DNA that are too small to analyze. In these cases, the scientist has to amplify the DNA using a lab method. This method is based on the process of DNA synthesis, which a cell uses to make a complete copy of DNA just before undergoing mitosis.

Geneticists may examine specific genes, or they may analyze the entire set of chromosomes in a patient. You may recall that human body cells contain 46 chromosomes arranged into 23 pairs. Scientists can isolate the chromosomes, stain them, and take a photo of the chromosomes as a group. They then can cut out images of individual chromosomes and organize them as part of a process known as karyotyping. By examining a person's karyotype, scientists can see if there are any extra or missing chromosomes. They also can determine whether any chromosomes are damaged.

To study specific genes, scientists begin by isolating DNA from cells. They cut the DNA into fragments and separate the fragments using a process known as electrophoresis. The separated DNA fragments then can be treated with a stain so that they can be seen. The pattern of fragments can help scientists compare DNA isolated from two people. The DNA then can be analyzed further to determine differences in their structures. A normal karyotype for a human contains 46 chromosomes, paired like this image.

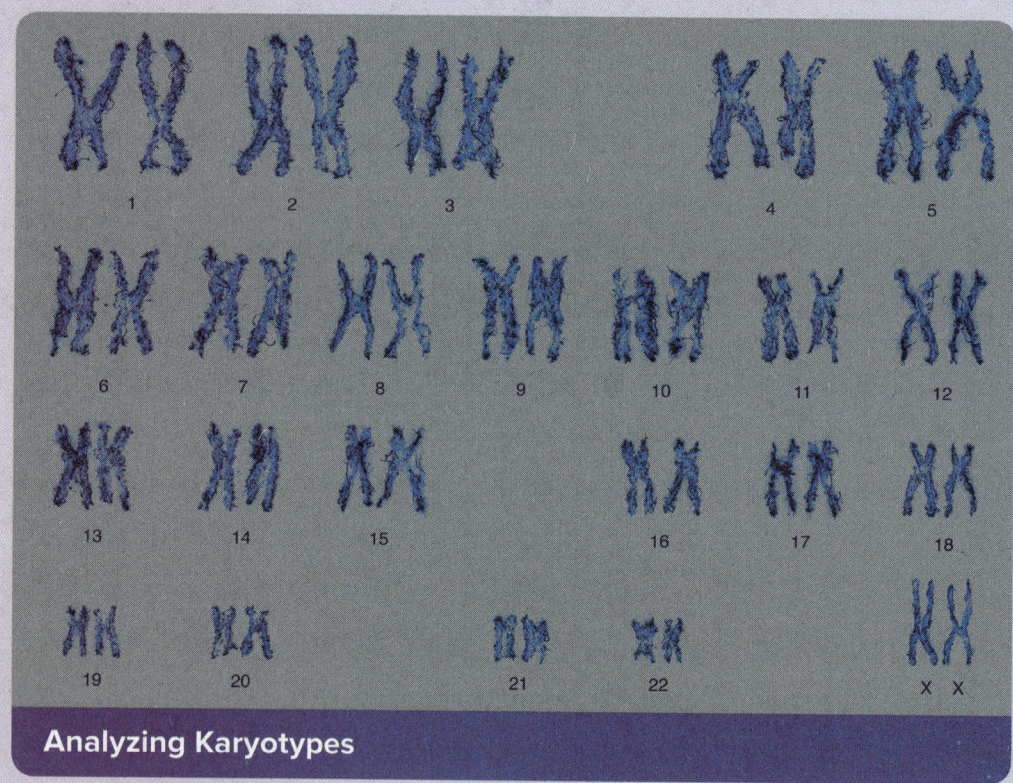

Analyzing Karyotypes

Karyotype Analysis

Conduct online research to learn about Down syndrome, Edwards syndrome, or Patau syndrome. Use what you learn to answer the questions that follow.

Name the syndrome you selected and **explain** how a karyotype analysis would reveal this syndrome in a patient.

_____
_____
_____
_____

Currently Available Treatments

What are some of the effects of the syndrome you selected? What are some of the currently available treatments?

_____
_____
_____
_____
_____
_____
_____

Gene Analysis

How may gene analysis or gene therapy help individuals with this disorder?

_____
_____
_____
_____
_____
_____
_____
_____
_____

ELABORATE

Concept 3.4: Heredity | 195

CONCEPT 3.4
How could a greater diversity of traits help a population adapt to a wider range of climates?

Activity 15
Concept Review

Review: Heredity

Quick Code
ca6594s

> Now that you have completed the objectives for this concept, review the core ideas you have learned. Record some of the core ideas below.

Core Ideas

_____
_____
_____
_____
_____
_____
_____
_____
_____
_____
_____
_____
_____
_____

Talk with a Group

Now, think about the annual average temperature map you saw in Get Started. Discuss how what you've learned about heredity can help you understand the annual average temperature map.

UNIT 3 | Unit Project

Solve Problems

Unit Project: Engineering a Better Banana

Does fruit grow this large in the wild?

Quick Code
ca6596s

Banana Tree

SEP Constructing Explanations and Designing Solutions
CCC Systems and System Models

You have learned about natural selection, the process of a species changing over time to survive in its environment. You also may be familiar with artificial selection. Humans use artificial selection to select favorable traits in an organism. Did you know that humans have been engineering bigger and better fruits and vegetables for many years?

Farmers who grow some types of fruit are at a disadvantage because they are not located where their fruit grows best. Tropical fruits, such as bananas, pineapples, mangoes, papayas, kiwis, and others, are grown in regions of the world known as tropical regions.

Tropical regions are the areas of the world near the equator that are characterized by warm weather with a moderate amount of rain all year long. These areas of the world exist because Earth is heated unequally by the sun, with the middle portion around the equator receiving most of the sunlight. Tropical regions are also located near large portions of the ocean, and these oceans are heated by the sun.

The heating of the ocean influences the weather patterns that produce the moderate rainfall. Looking at a globe or map of the world, tropical regions can be identified as those regions that are located in latitudes near the equator.

Engineering and Growing Corn

Unit Project

Maximizing Yield

What are some technologies used to maximize corn yields? What are some side effects of using these technologies?

Where Are the Bananas?

Study the climate map below. Using the key to the left of the map, **write** an X on the zone or zones where you would expect to find bananas growing.

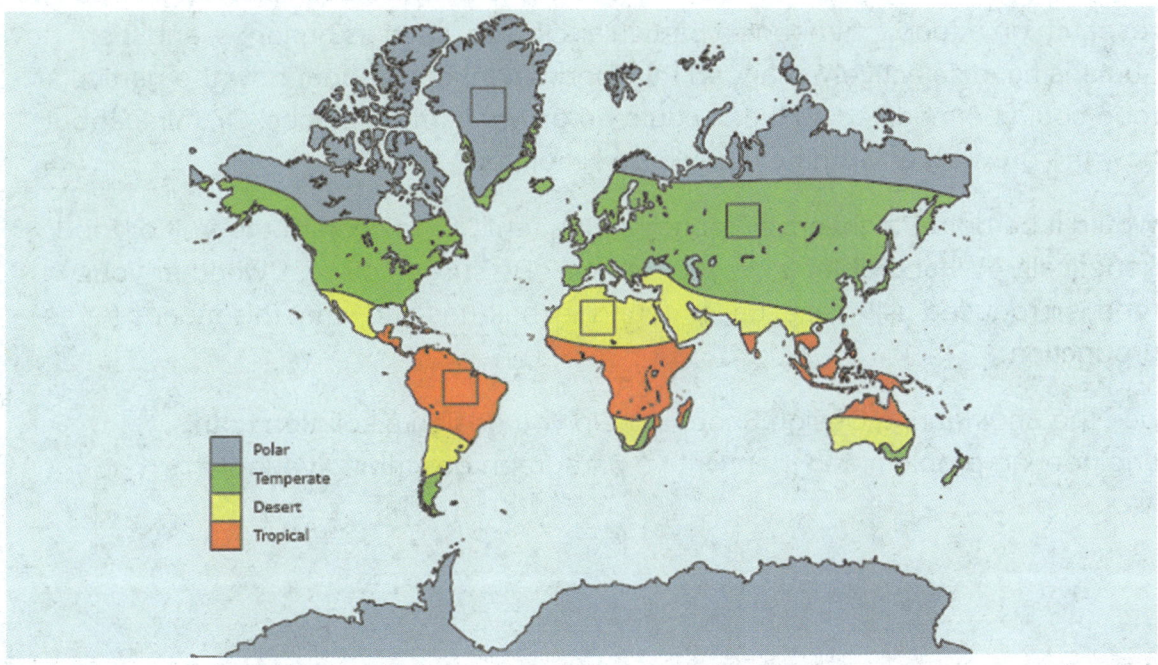

Unit Project

Modified Foods

Many foods are produced by artificial breeding or genetic modification. Some foods, such as bananas, are produced using asexual techniques. Other foods, such as corn, are produced using hybrid techniques. Other bioengineering techniques are also used. Some of these techniques need a particular climate. Conduct online research on a food that needs a particular climate, such as bananas, and learn how humans have selectively changed this food. Then, think about how the particular food could be modified so that it could be grown in more places. Or, think about how the growing environment could be simulated.

Would it be better to invent a "food growing tent," or to change the fruit or food genetically? What problems do you foresee? Be sure to discuss whether your food is produced using asexual or hybrid techniques and what this means for its production.

Use the Engineering Design Sheet to help you design a solution to this engineering problem and then write a paragraph describing your plan.

Student Engineering Design Sheet

- What is the problem?
- What ideas do you have?
- How will you pick one idea to test?

Ask About the Problem

Design and Plan

- How will you know your idea works?
- What worked? What did not work?

Improve

- What could work better?
- What is your final idea?

UNIT 4

Our Changing Climate

Unit 4: Our Changing Climate | 205

UNIT 4 | Get Started

Measuring Climate Change

This map shows the difference in temperature compared to the average recorded temperatures between 1981 and 2010. How has the temperature changed where you live?

Quick Code
ca6751s

Guiding Questions

1. How do human activities affect Earth's systems?

2. How do we know our global climate is changing?

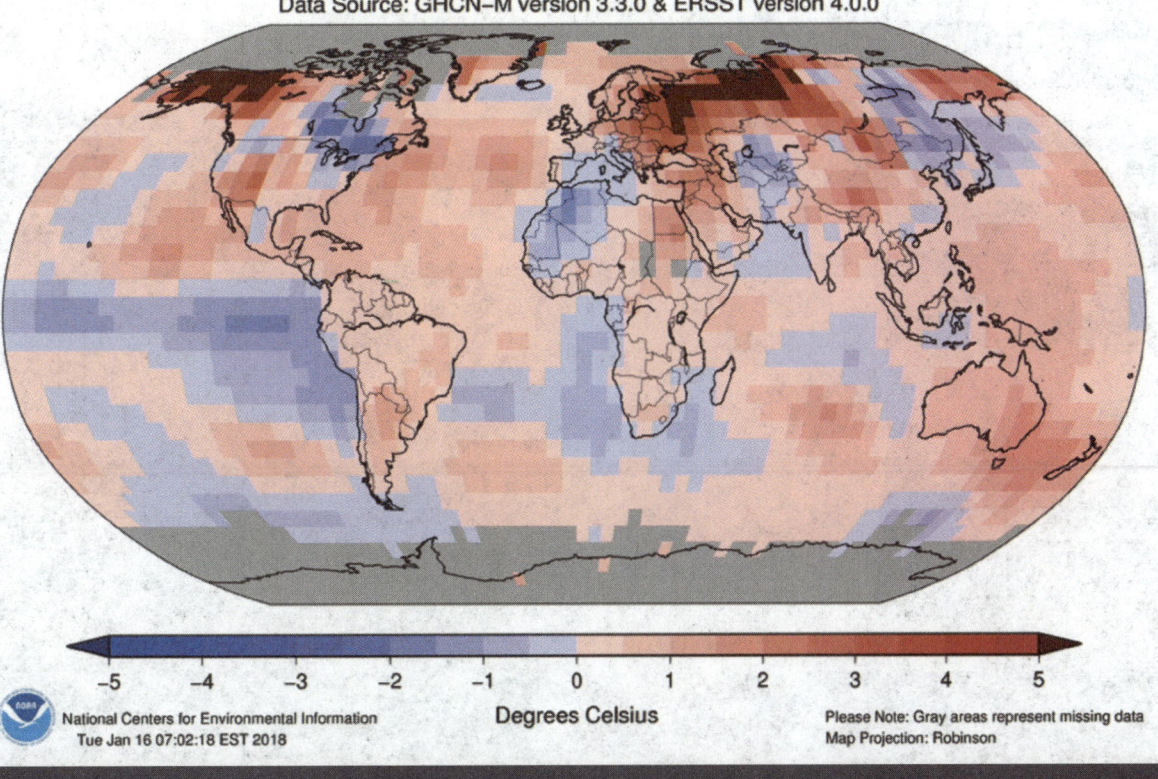

Measuring Climate Change

Unit Project Preview

Solve Problems

STEM

Unit Project: Cow Pollution

Thinking About Solutions

Do cows contribute to global warming?

You are part of a team of scientists interested in reducing the impact of cows' methane production. Before moving forward, determine what you already know, what you need to learn, and how to find this data. Consider what strategy you might use, as well as ethical considerations and any technical difficulties you may encounter.

Quick Code ca6750s

Hereford Cow

Use the graphic organizer to **sort** your ideas.

What I Know

What I Need to Know

How Can I Find Out?

CONCEPT 4.1

Causes of Climate Change

Student Objectives

By the end of this lesson:

- [ ] I can synthesize information to describe and predict how many natural and human-induced factors interact to cause a rise in global temperatures on Earth over time.

- [ ] I can develop models that demonstrate how change in one of Earth's systems can cause change in one or more of Earth's systems and how, over time, changes disrupt Earth's equilibrium.

- [ ] I can develop a model that describes and predicts the effects of greenhouse gases on Earth's temperature.

- [ ] I can evaluate data to explain how Earth's climate has changed over time and to calculate the rates of these changes.

- [ ] I can argue from evidence that increased greenhouse gas emissions cause changes in global sea levels and global land and ocean temperatures.

Key Vocabulary

anthropogenic, atmosphere, climate, data, drought, El Niño, fossil, fuel, global warming, greenhouse gases, hemisphere, ice core, latitude, meteorite, radiation, sunspot, topography, weather

Quick Code
ca6753s

Activity 1

Can You Explain?

Why don't Earth's natural processes account for the observed global climate change?

How can we determine what is causing the change?

Quick Code
ca6754s

Concept 4.1: Causes of Climate Change

CONCEPT 4.1 — Why don't Earth's natural processes account for the observed global climate change? How can we determine what is causing the change?

Activity 2
Ask Questions

Climate Change Models

Quick Code
ca6755s

Look closely at the graph.

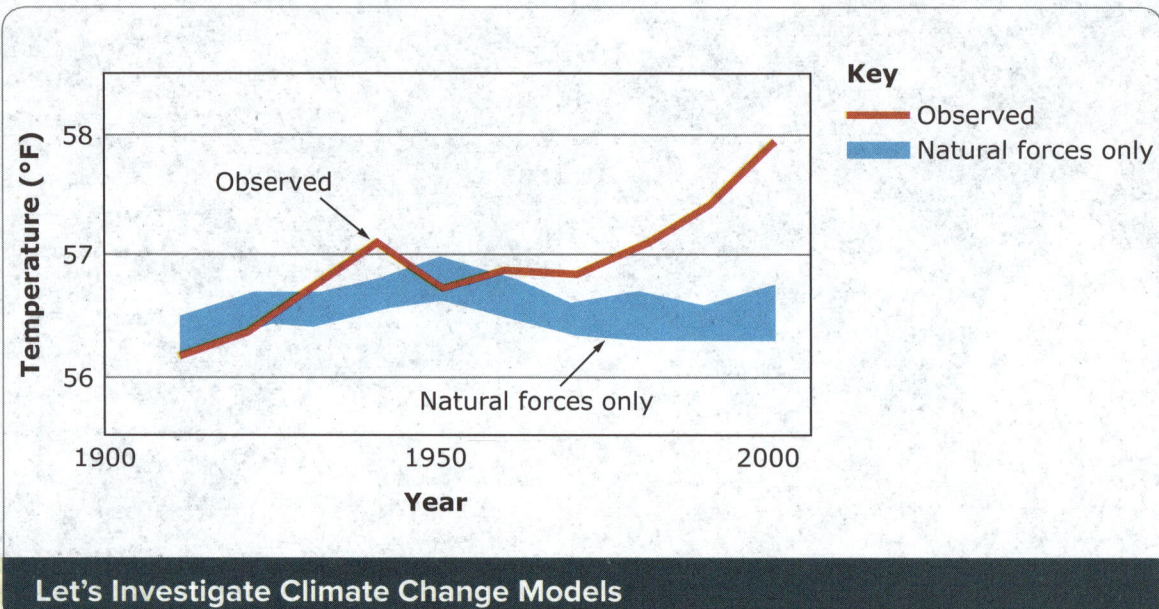

Let's Investigate Climate Change Models

Brainstorm questions that you have after looking at this graph.

_____

_____

_____

_____

_____

_____

SEP Asking Questions and Defining Problems

SEP Analyzing and Interpreting Data

Categorize your questions. For example, are they research questions? Experimental? Opinion?

Concept 4.1: Causes of Climate Change

CONCEPT 4.1 Why don't Earth's natural processes account for the observed global climate change? How can we determine what is causing the change?

Activity 3
Observe

Inside a Glacier and Dinosaurs in the Badlands

Quick Code
ca6756s

Watch one of these videos: Inside a Glacier or Dinosaurs in the Badlands. Then, **answer** the questions that follow.

Inside a Glacier

Dinosaurs in the Badlands

SEP Obtaining, Evaluating, and Communicating Information
CCC Stability and Change

Based on the video you saw, perform the following activities.
Draw a diagram that shows the layers of a glacier or the layers in a rock formation and how they relate to time. **Share** your diagram with your assigned group.

CONCEPT 4.1 | Why don't Earth's natural processes account for the observed global climate change? How can we determine what is causing the change?

Discuss with your group how scientists could use information from the layers as data or evidence in a scientific investigation. **Note** your group's ideas for this below.

_____
_____
_____
_____
_____
_____
_____
_____
_____
_____

Activity 4
Evaluate

Quick Code
ca6757s

What Do You Already Know About Causes of Climate Change?

What Causes Climate Change?

Which kinds of events directly cause Earth's climate to change? **Check** all that apply.

- ☐ volcanic eruptions
- ☐ earthquakes
- ☐ sunspot activity
- ☐ large meteorite impacts
- ☐ hurricanes

CONCEPT 4.1

Why don't Earth's natural processes account for the observed global climate change? How can we determine what is causing the change?

What Causes Seasons?

Which statements about Earth are true? **Check** all that apply.

☐ Seasonal changes are the result of changes in the distance between Earth and the sun.

☐ Earth's climate is always the same.

☐ Seasonal changes are the result of the tilt of Earth's axis as Earth travels around the sun.

☐ The original source of heat energy in Earth's atmosphere is the sun, and solar energy radiates through space to reach Earth's surface.

How Does Earth Orbit?

Which statements about Earth's orbit around the sun are true? **Check** all that apply.

☐ Earth's orbit changes shape over time.

☐ Earth's orbit is extremely elliptical in shape.

☐ Earth's orbit is perfectly circular in shape.

☐ Earth's orbit is almost circular but is slightly elliptical.

Climate-Changing Activities

Each item below describes daily activities.

Classify each activity as having greater, smaller, or about the same net contribution to climate change. **Place a checkmark** in the appropriate column.

Daily Activity	Greater	Smaller	About the Same
Eating a burger at a fast food joint			
Eating salad at home			
Buying new clothes made in another country			
Patching a pair of jeans to use them again			
Coming to school in a motor vehicle			
Cycling to school			
Using plastic bags to carry groceries home			
Using bags made from recycled paper to carry groceries home			
Drying clothes in an electric dryer			
Drying clothes on a clothesline outside			

Concept 4.1: Causes of Climate Change

How Do Natural Processes Affect Global Temperature Change?

Activity 5
Analyze

Quick Code ca6761s

Natural Climate Changers

Read the following text. As you read, **highlight** information you can use as evidence to support your initial ideas on how to answer the Can You Explain? question or the questions you generated during Engage. **Record** each natural process that causes climate change in the chart at the end. Then, **record** whether you think each change would increase or decrease global temperatures.

Natural Climate Changers

The natural process of the sun's radiation heating Earth creates our weather. Earth's climate depends on how much radiation the sun emits, how much reaches Earth's surface, and how the heat is transported around the globe by oceans and the atmosphere. Changes to the amount of radiation Earth receives can cause changes in climate. Short-term changes occur over periods of years, perhaps a decade. Long-term changes occur over periods of tens of thousands, or hundreds of thousands, of years.

CCC Stability and Change

The amount of radiation the sun emits can be influenced by sunspots and solar flares. Sunspots look like dark spots on the sun, and the average sunspot is the same diameter as Earth. The dark spots are caused by cooler sections of the sun's surface that eventually erupt. These eruptions—called solar flares—release huge amounts of gas, heat, and energy into space. Scientists have been tracking sunspot activity for centuries and have found two key pieces of information. First, sunspot activity happens in 11-year cycles, with times when very few sunspots can be seen and times when many sunspots are visible. Second, periods of low sunspot activity correlate with cooler periods of Earth's climate. The connection between sunspot activity and Earth's climate is still not completely understood, and many scientists are working to learn more about the phenomenon.

The cycles of El Niño and La Niña are much better understood than sunspots. In a normal year, the cold water in the South Pacific rises to the surface, and the trade winds blow westward. During an El Niño year, the cold water does not rise, which leads to warmer surface water. The warmer water makes the trade winds weaker, so areas east of the South Pacific will see rainstorms and flooding. Areas west of the South Pacific will experience drought. Interestingly, the northern and central United States will actually have a warmer winter as a result of El Niño. During a La Niña year, ocean temperatures are cooler than usual. This results in the opposite conditions of El Niño, with flooding west of the South Pacific and droughts to the east. Although El Niño and La Niña occur in cycles, it can be difficult to predict when exactly they will occur.

Natural Climate Changers *cont'd*

Beyond the initial devastating effects of large volcanic eruptions, volcanoes also can cause short-term changes to the regional and global climates. When a large volcano erupts, it releases a great deal of ash and gas into the atmosphere. The dust and gas circulate in Earth's atmosphere, carried by the global wind currents. This material can block some of the sun's radiation from reaching Earth's surface, resulting in cooler temperatures locally. The cooler temperatures can be global if the eruption is large enough.

Large meteorite strikes have an effect on the climate that is similar to the effect of volcanic eruptions. The initial impact can send ash and dust high into the atmosphere to circulate and block out the sun's radiation, resulting in cooling. A meteorite large enough to change the climate would completely devastate the surrounding area with its impact. It would send out shock waves that flatten trees and debris that kills on impact. Scientists believe that a meteorite may have been responsible for the extinction of the dinosaurs. Life in the immediate area would have been destroyed, and life worldwide would have felt the effects of global cooling.

In the following chart, **record** each natural process that changes Earth's climate. Then, **write** whether it would change the climate by increasing or decreasing global temperatures.

Natural Process That Causes Climate Change	Increase or Decrease Temperatures?

Concept 4.1: Causes of Climate Change | 225

CONCEPT 4.1 Why don't Earth's natural processes account for the observed global climate change? How can we determine what is causing the change?

Activity 6
Observe

Mount Pinatubo Eruption

Quick Code
ca6762s

Study the image of the Mount Pinatubo eruption. Then, **respond** to the following questions.

Mount Pinatubo Eruption

Make observations that support the claim that the eruption released material into the atmosphere.

_____

_____

_____

_____

_____

_____

What materials were released into the atmosphere that could affect climate?

_____

_____

_____

_____

_____

_____

_____

CONCEPT 4.1 | Why don't Earth's natural processes account for the observed global climate change? How can we determine what is causing the change?

Study the graph Effect of Mt. Pinatubo Eruption, and then answer the following questions.

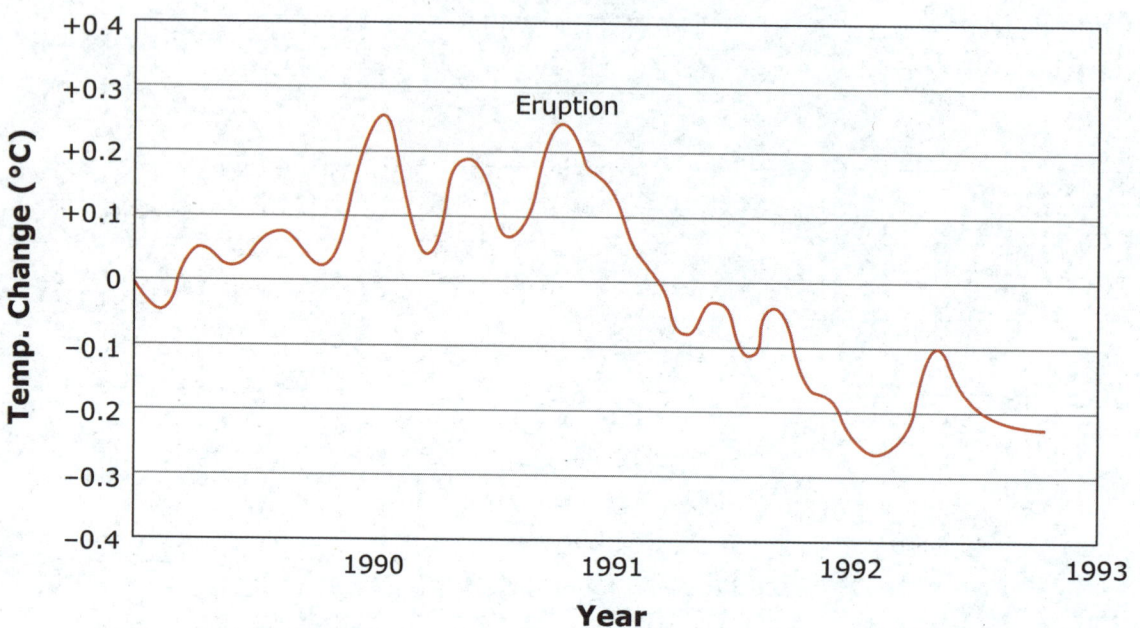

What happens to the temperature after the eruption occurs?

What is the total change in temperature from the time of the eruption to the lowest temperature point?

CONCEPT 4.1 | Why don't Earth's natural processes account for the observed global climate change? How can we determine what is causing the change?

Activity 7
Investigate

Hands-On Investigation: Model a Meteorite Impact

Quick Code
ca6763s

In this investigation, you will model a meteorite impact.

Predict

Does a meteorite's impact affect Earth and the atmosphere?

_____
_____
_____
_____
_____
_____
_____
_____
_____
_____
_____
_____

What materials do you need? (per group)

- Clay powder
- Aluminum foil pan, 13×9×2
- Construction paper
- White crayon
- Metric ruler
- Marbles, $\frac{5}{8}$ in.
- Masking tape
- Safety goggles (per student)

Procedure

1. Wear your safety glasses throughout this investigation.

2. With a partner, fill a foil pan with dry, loose clay to a depth of 1 inch.

3. Prepare your piece of construction paper for the activity by placing a ruler in the middle of the paper along a vertical line. Use a white crayon and reproduce the scale of the ruler (in quarter-inches or half-centimeters, depending on the ruler) along the paper. Place the ruler in the middle of the paper along a horizontal line, and use the white crayon to reproduce the ruler's scale again. The result should be two sets of "rulers" crossing perpendicularly in the middle of the construction paper.

4. Hold or tape the prepared piece of construction paper behind the tray so that its two ruled scales are visible.

5. Decide with your partner which of you will observe the tray at eye level and which will drop the marble into the tray. Drop the marble into the tray from a height of about 18 inches. Then, switch roles with your partner and repeat dropping the marble. Both you and your partner should use the ruled scale on the black piece of paper to approximate the height and width of the plume that results when the marble hits the clay surface. Record this data in the space below.

Trial	Height of Plume	Width of Plume

Concept 4.1: Causes of Climate Change

CONCEPT 4.1

Why don't Earth's natural processes account for the observed global climate change? How can we determine what is causing the change?

Reflect

Describe the patterns you observed upon impact and shortly thereafter.

_____
_____
_____
_____
_____
_____
_____

What are some limitations of this model of a meteorite impact?

_____
_____
_____
_____
_____
_____
_____

Compare the effects of volcanic eruptions that you analyzed in the Mount Pinatubo Eruption activity and the meteorite impacts explored in the Model a Meteorite Impact activity. **Record** your thoughts in the Venn diagram below.

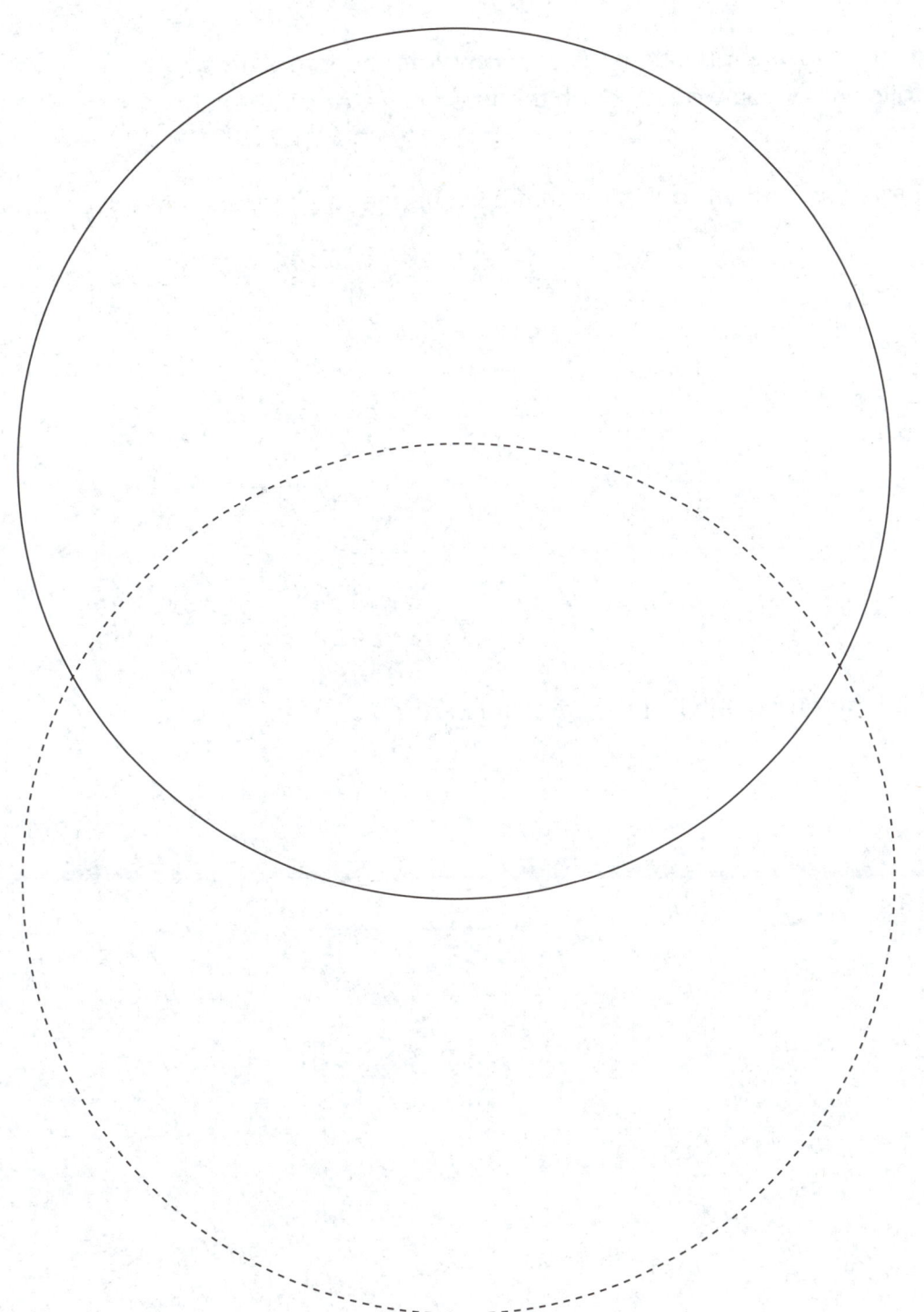

CONCEPT 4.1 | Why don't Earth's natural processes account for the observed global climate change? How can we determine what is causing the change?

Activity 8
Observe

Our Climate in the Long Run

Quick Code
ca6764s

Read the following questions. Then, **complete** the Interactive that follows. As you work through the Interactive, **answer** each question.

What are two Earth irregularities that can change the intensity of the seasons?

_____
_____
_____
_____
_____

How did the end of the last ice age cool Earth?

_____
_____
_____
_____
_____
_____

Why are scientists concerned about increased jet exhaust?

Go online to explore Earth's climate and what impacts it.

Our Climate in the Long Run

Activity 9
Analyze

Earth's Orbits

Quick Code
ca6765s

Read the text. As you read, **look** for possible causes of the periodic ice ages experienced on Earth.

Earth's Orbits

The spherical shape of Earth, the tilt of its axis, and its orbit around the sun cause seasons. We experience summer when the Northern **Hemisphere** is tilted toward the sun, thus receiving more direct solar **radiation**. We experience winter when the Northern Hemisphere is tilted away from the sun, and the heat energy the sun radiates is spread over a larger area.

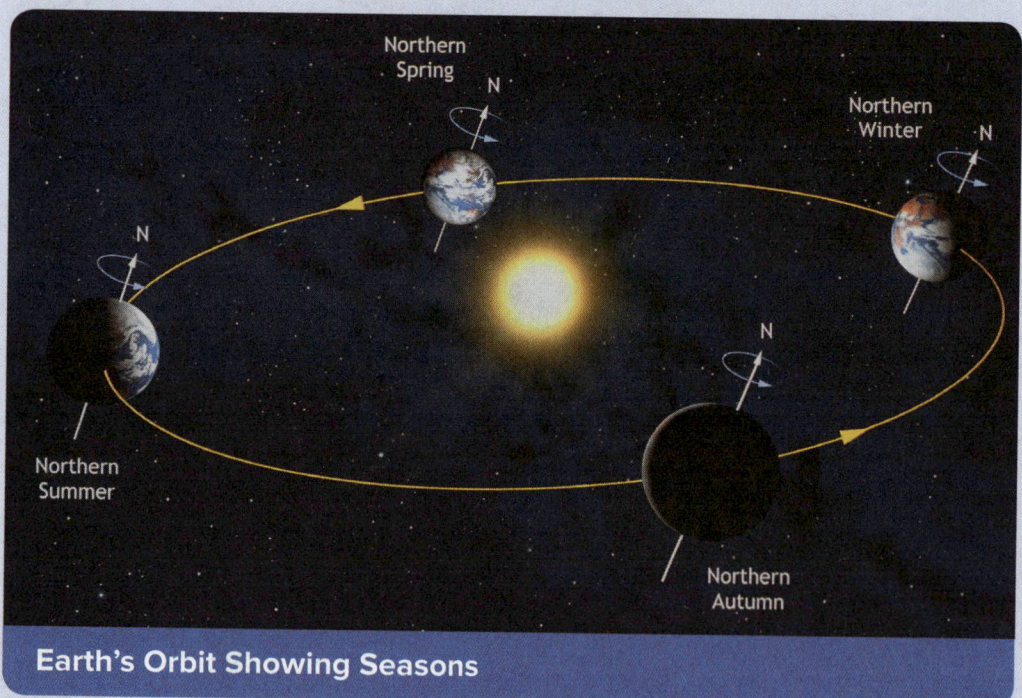

Earth's Orbit Showing Seasons

Many people are unaware that Earth wobbles on its axis. At times, Earth's axial tilt is more extreme; at other times, it is less extreme. When Earth's axial tilt is less extreme, summers and winters are milder. Variations in Earth's axial tilt occur over a period of about 41,000 years. Earth's orbital path also changes in a cyclical pattern.

The shape of Earth's orbit around the sun is an ellipse. However, sometimes this ellipse is almost perfectly circular, and sometimes it is more elongated. This variation happens in cycles that take hundreds of thousands of years. When Earth's orbit is more elliptical, seasonal differences are more extreme. When Earth's orbit is more circular, seasonal differences are less extreme. The shape of Earth's orbit is influenced by gravitational interactions with Jupiter and Saturn. Together, they can pull Earth's orbit away from being generally circular. These interactions do not change the length of Earth's year.

Although we know some of the external factors that affect Earth's long-term **climate**, it appears that changes in Earth's orbit and rotation can account for only some of the changes in temperature. There are also feedback mechanisms on Earth that influence climate. Sometimes, a slightly colder climate can contribute to an even colder climate. Glaciers are a primary source of this.

Video
Periodic Ice Age

Earth's Orbits *cont'd*

Recall that light can be reflected, and it is well reflected, by white surfaces. Glaciers have white, icy surfaces. Hence, they reflect a lot more sunlight than the surface of Earth normally would if it were only rock and vegetation. Less sunlight being absorbed means lower temperatures locally. With enough glaciers covering Earth, they can contribute to lower temperatures globally.

Iceberg Fragment

What are possible causes of the periodic ice ages experienced on Earth? Use evidence from the text to support your answer.

CONCEPT 4.1 Why don't Earth's natural processes account for the observed global climate change? How can we determine what is causing the change?

Activity 10
Evaluate

La Niña and Sunspots and Tilts and Orbits

Quick Code
ca6766s

La Niña and Sunspots

La Niña, sunspots, and solar flares can all have a short-term effect on climate. During a particular year, scientists made the connection that the La Niña was occurring during a time when there were fewer sunspots and solar flares. **Fill in the blanks** to create a true statement about their effects on southern California.

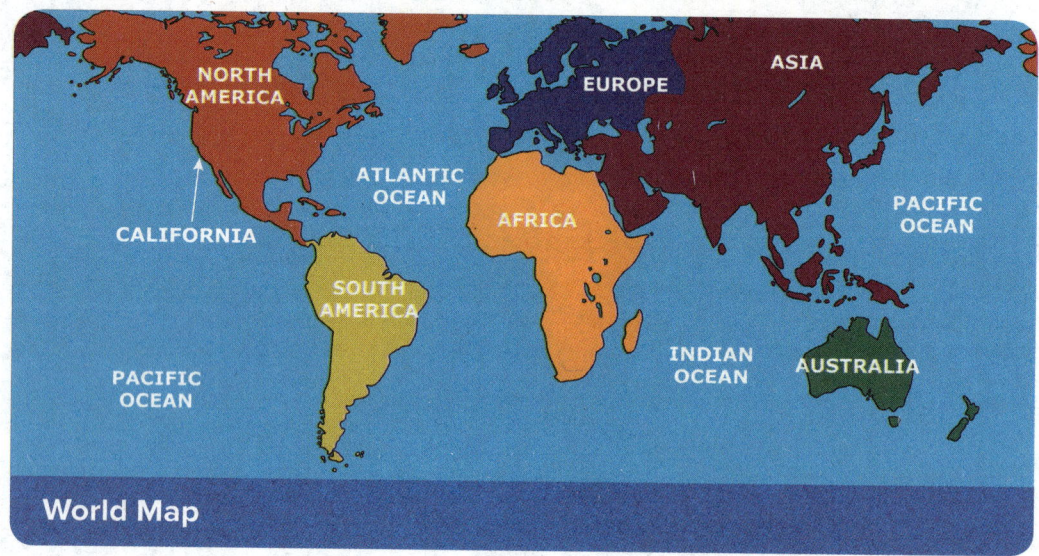

World Map

La Niña's _____ [**stronger/weaker**] trade winds would cause _____ [**cooler/warmer**] than normal ocean temperatures for southern California, and it would experience _____ [**flooding/drought**] during the winter. During a cycle of fewer sunspots and solar flares, temperatures tend to be _____ [**lower/higher**]. One would predict that southern California's winter would be _____ [**warmer/cooler**] and _____ [**wetter/drier**] than normal.

CCC Cause and Effect

Tilts and Orbits

The following text explains how the movement of Earth affects long-term climate change.

Highlight the errors in the text, and then **rewrite** the sentences to correct the errors.

Earth's climate is strongly affected by the amount of solar energy Earth receives. Several factors affect long-term climate change on Earth. Long-term climate change occurs over periods of hundreds of years. The shape of Earth's orbit affects the amount of solar energy Earth receives. When Earth's orbit is nearly circular, the amount of energy received is much more consistent, whereas when Earth's orbit is more elliptical, the amount of solar energy received is more varied. In the winter and during ice ages, the sun's gravity is not as strong, so Earth is farther away from the sun, which leads to significantly less solar energy received. Changes in Earth's tilt also affect the long-term climate of Earth. When the tilt is greater, the sun's energy is spread more evenly, and the seasons are less pronounced, with summers being less different from winters.

_____
_____
_____
_____
_____
_____

How Does Human Activity Affect Global Temperature Change?

Activity 11
Analyze

Human Impact

Quick Code
ca6767s

Read the text and **answer** the following questions. Then, **write** down additional questions you would like to explore.

Human Impact

You know that Earth's climate changes due to natural processes like sunspots, volcanic eruptions, and the slight changing of Earth's axis and orbit. But you have also seen a graph that shows that the modeled climate change due to natural processes does not account for the observed increase in global temperature averages. The graph that follows shows another set of data.

Humans have always made an impact on the environment because humans rely on the environment in so many ways. As global populations have grown over time, the impact has likewise grown. Since the middle of the 1800s, when large-scale industry began, human activities have had a substantial warming effect on Earth's climate. This human impact greatly exceeds impacts due to changes from natural processes, like volcanic eruptions or cycles in the sun's output of energy.

CCC Stability and Change

Temperature Increases

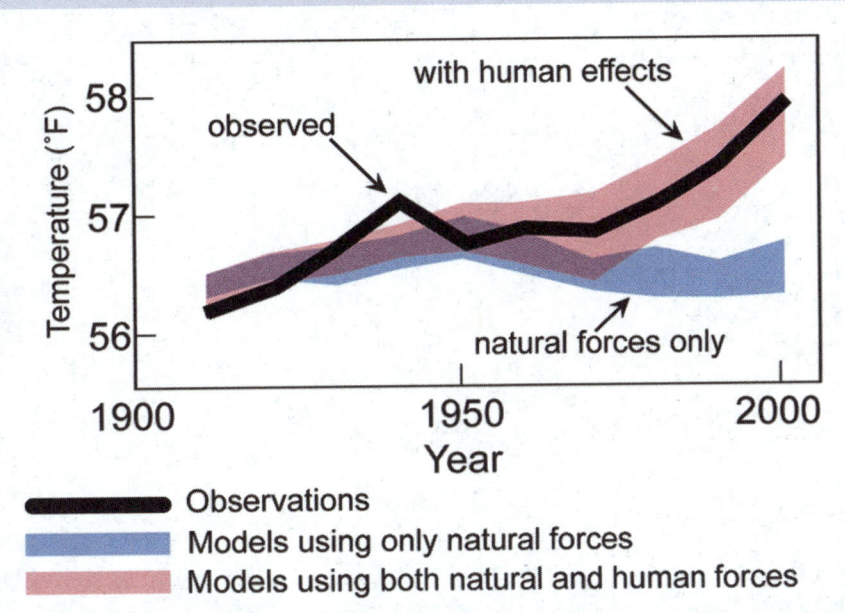

The biggest **anthropogenic**, or human-related, cause of **global warming** and climate change is the burning of **fossil fuels** such as coal, oil, and natural gas.

These fuels contain high-energy substances called hydrocarbons. Fossil **fuel** burning releases carbon dioxide, water vapor, and other **greenhouse gases** into the **atmosphere**, which leads to the greenhouse effect and global climate change.

Is a temperature change of a few degrees a lot or not a lot?

_____
_____
_____
_____
_____
_____
_____

Is 100 years a long time or not a long time?

_____
_____
_____
_____
_____
_____
_____

Are the answers to these questions always the same?

___

What additional questions do you have about the text?

___

CONCEPT 4.1 — Why don't Earth's natural processes account for the observed global climate change? How can we determine what is causing the change?

Activity 12
Interpret Data

Observe the Greenhouse Effect

Quick Code
ca6768s

In this investigation, you will use a covered container to observe the greenhouse effect. Then, you will describe the theory behind global warming and the possible consequences of the greenhouse effect on Earth.

Predict

Does the greenhouse effect affect life on Earth?

_____

_____

_____

_____

_____

_____

_____

_____

_____

_____

_____

SEP Developing and Using Models

What materials do you need? (per group)

- Clear terrarium lids
- Plastic container, 2 qt
- Thermometers

Procedure

Place a clear container in the sunlight with the lid off. Place the thermometer in the bottom of the container. Wait 5 minutes. Record the temperature reading in the data table below. Using the second container, repeat the steps with the lid on.

Temperature with Lid On	Temperature with Lid Off

Answer the following questions.

Why did the air temperature inside the container increase with the lid on?

_____

_____

_____

How does the terrarium greenhouse compare to the greenhouse effect in Earth's atmosphere?

_____

_____

_____

_____

Concept 4.1: Causes of Climate Change

CONCEPT 4.1

Why don't Earth's natural processes account for the observed global climate change? How can we determine what is causing the change?

Reflect

How does the covered container model our atmosphere? What are the limitations of this model?

_____
_____
_____
_____
_____
_____
_____

Explain why the term *greenhouse effect* is used to describe the theory of global warming.

_____
_____
_____
_____
_____
_____
_____
_____

What are the possible effects of a buildup of greenhouse gases in our atmosphere?

___

___

___

___

___

___

___

___

CONCEPT 4.1 | Why don't Earth's natural processes account for the observed global climate change? How can we determine what is causing the change?

Activity 13
Observe

The Greenhouse Effect

Quick Code
ca6769s

As you **watch** the video, **write** down what you have learned and any questions you may have.

The Greenhouse Effect

___

___

___

___

___

___

___

SEP Obtaining, Evaluating, and Communicating Information

Activity 14
Interpret Data

Quick Code
ca6770s

World Energy Use

This table shows data from 1980 to 2008. The first data column shows the population of a region of China each year. The second data column shows energy consumption, per person, for each of those years. **Calculate and graph** the region's total energy consumption for each year. Be prepared to compare graphs and discuss how the pattern may affect greenhouse gases in the atmosphere.

Energy Use in a Region of China

Year	Approximate Population (thousands of people)	Approximate per Capita Energy Consumption (million BTU per person)
1980	100	20
1984	110	20
1988	110	25
1992	120	25
1996	120	30
2000	130	30
2004	130	45
2008	130	60

SEP Using Mathematics and Computational Thinking

CONCEPT 4.1
Why don't Earth's natural processes account for the observed global climate change? How can we determine what is causing the change?

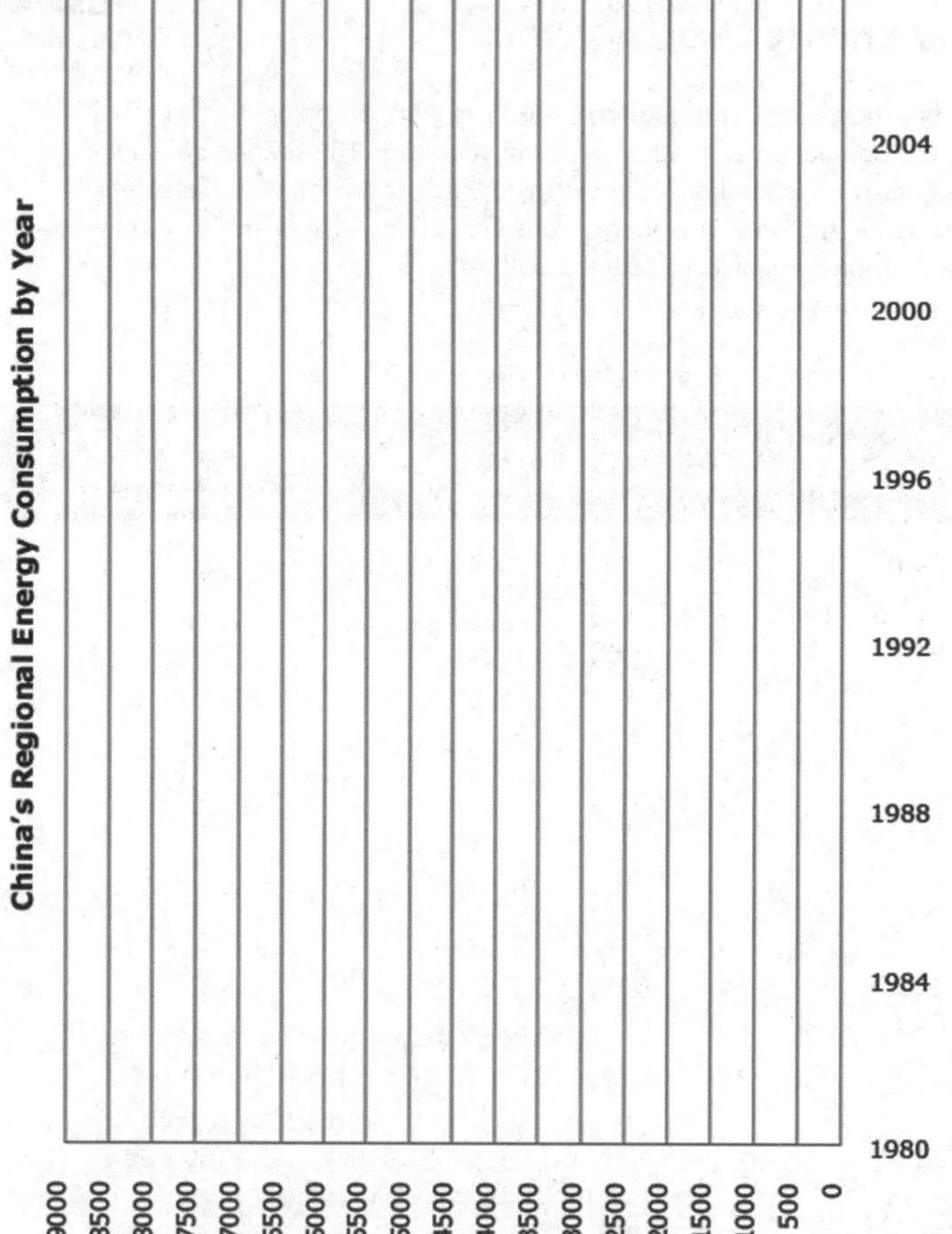

Activity 15
Evaluate

Evidence of Climate Change

Quick Code
ca6771s

Study the graphs. Then, **read** each of the questions. **Circle** the question that cannot be answered by the graphs.

Evidence of Climate Change

(a) Global Land and Ocean Temperature Changes

(b) Global Sea Level Change

(c) Global Greenhouse Gas Concentrations

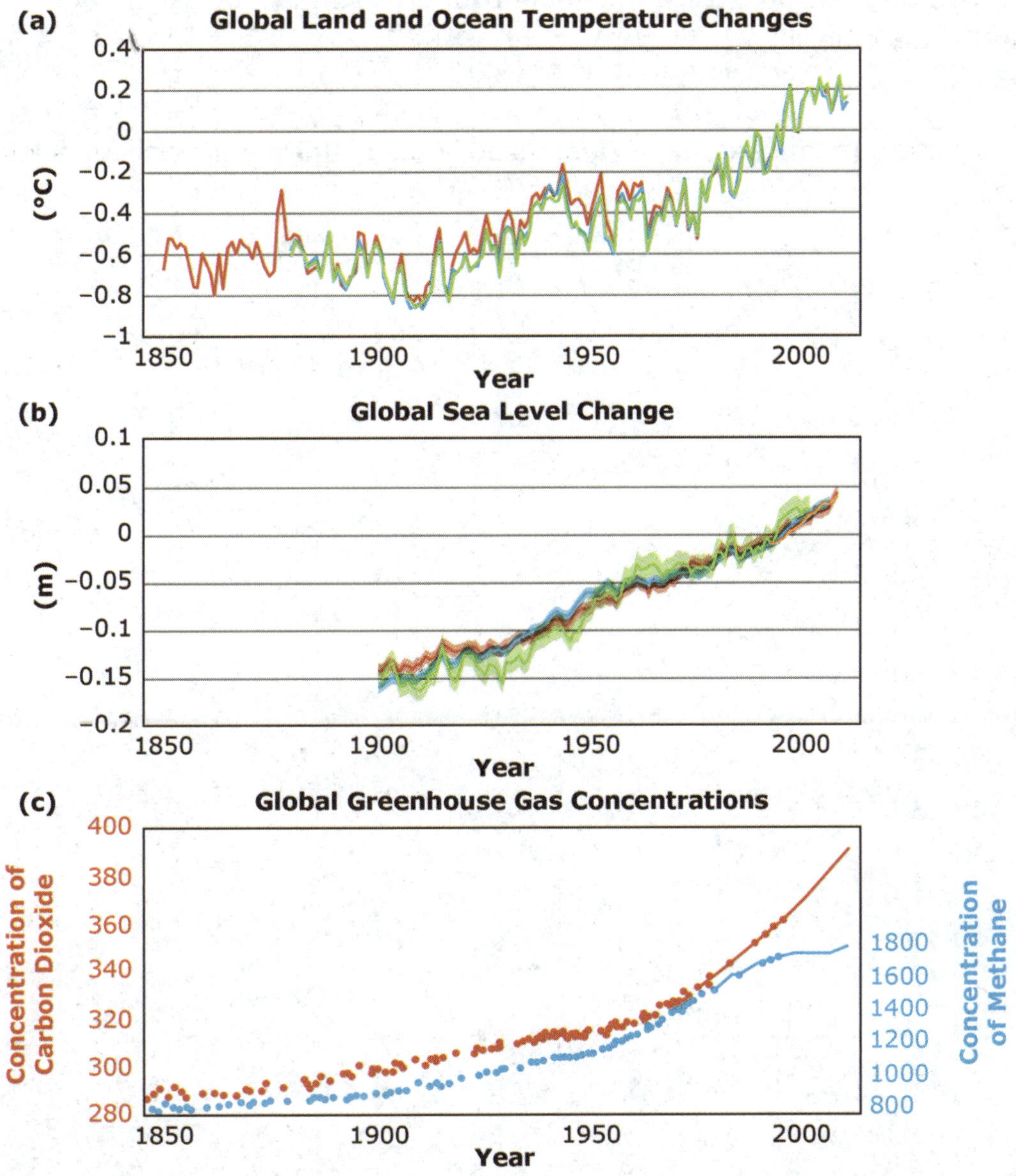

SEP   Analyzing and Interpreting Data

Concept 4.1: Causes of Climate Change | 253

CONCEPT 4.1 Why don't Earth's natural processes account for the observed global climate change? How can we determine what is causing the change?

How do increasing land and ocean temperatures affect sea level?

Are greenhouse gases in the atmosphere related to land and ocean surface temperature?

If greenhouse gas concentrations continue to rise, will the sea level also continue to rise?

Does increasing carbon dioxide in the atmosphere increase the likelihood that people will become ill?

If greenhouse gas concentrations decreased, would temperatures and sea levels decrease?

What Is the Evidence for Global Warming?

Activity 16
Analyze

Quick Code ca6772s

Data about Global Warming

Read the text. As you read, **highlight** the things scientists can directly measure.

Data about Global Warming

Scientists have done a great deal of research to determine the precise nature and causes of this change in Earth's climate. Scientists have documented the increase in atmospheric carbon dioxide by sampling the atmosphere and analyzing its composition. Direct sampling shows that carbon dioxide levels in the atmosphere have increased steadily over the past 60 years. These data represent only a very short period of Earth's history.

Additional data comes from the study of ice cores taken from Earth's poles. An ice core is a sample of ice taken from an ice sheet; the data can go back as far as 400,000 years. Because ice cores contain small bubbles of trapped air, they can serve as a record of the concentration of gases in Earth's atmosphere.

Science Nation: Greenland Ice Cores

Concept 4.1: Causes of Climate Change | 255

Data about Global Warming *cont'd*

Ice cores help scientists understand how Earth's climate has changed over hundreds of thousands of years. The evidence from ice cores verifies that concentrations of atmospheric carbon dioxide are the highest they have been in human existence.

Sometimes, people may doubt that global climate change is happening, but more than 1,300 scientific experts at the Intergovernmental Panel on Climate Change (IPCC) collaborate and share data to provide evidence that human activities are causing rapid climate change.

Increased carbon dioxide in the atmosphere and air bubbles in ice sheets can be abstract ideas. Because we cannot physically see this evidence, it is difficult to understand what it means or why it matters. Other evidence is easier to see and understand, and it may help prove that Earth is warming due to increased greenhouse gases produced by humans. For example:

- Ice sheets in Antarctica and Greenland have shrunk noticeably as the ice has melted. Some of the water released has contributed to a rise in sea level. The level of the seas around the world has risen about 17 centimeters in the last 100 years.

- As ocean temperatures have warmed, the ocean water has expanded. This has also contributed to a rise in sea levels.

- The area covered by sea ice has declined rapidly over the past few decades.

- Glaciers on mountains around the world are melting.

- Snow cover has declined, and winter snow is melting faster.

- Previously permanently frozen ground, permafrost, is melting in some tundra areas.

- Oceans are becoming more acidic. This acidification has occurred because the oceans are absorbing more carbon dioxide from the atmosphere.

- Extreme **weather** events are becoming more frequent, such as the increasing number of record high temperature events in the United States.

Create a shape poem in the shape of the sun using the terms from the following list:

- Atmospheric carbon dioxide
- By sampling the atmosphere and analyzing its composition
- An ice core
- Concentration of gases in Earth's atmosphere
- Ice sheets in Antarctica and Greenland
- The seas around the world have risen about 17 cm in the last 100 years
- Ocean temperatures have warmed
- Rise in sea levels
- Area covered by sea ice
- Snow cover
- Oceans becoming more acidic
- Extreme weather events are becoming more frequent

Activity 17
Interpret Data

Global Temperatures and Carbon Dioxide

Quick Code
ca6773s

Analyze the following graphs. Then, **write** a brief comparison of the two graphs and a statement about how carbon dioxide levels in the atmosphere may affect global temperatures.

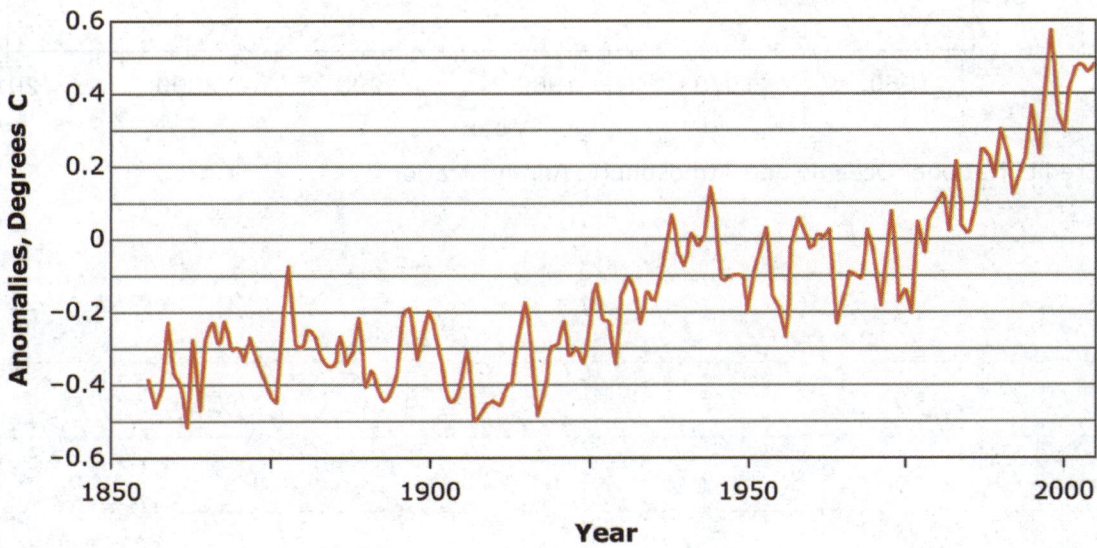

Credit: NCDC and NOAA

SEP Analyzing and Interpreting Data

Concept 4.1: Causes of Climate Change | 259

CONCEPT 4.1 — Why don't Earth's natural processes account for the observed global climate change? How can we determine what is causing the change?

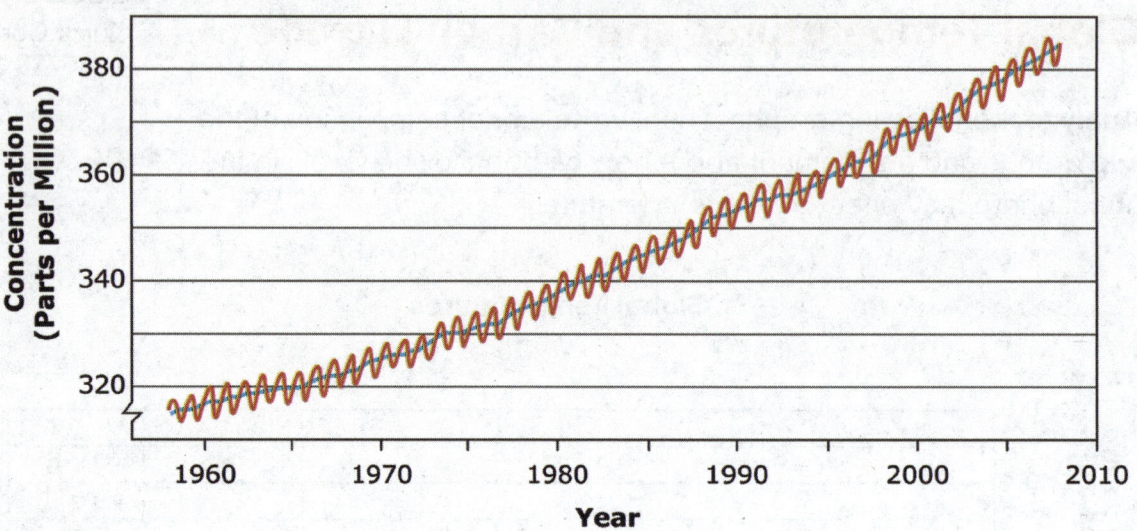

Credit: National Oceanic and Atmospheric Administration

Activity 18
Evaluate

Global Temperature Change

Quick Code
ca6774s

Study the line graph. **Complete** the following sentences by interpreting the graphical data and using your knowledge of global warming.

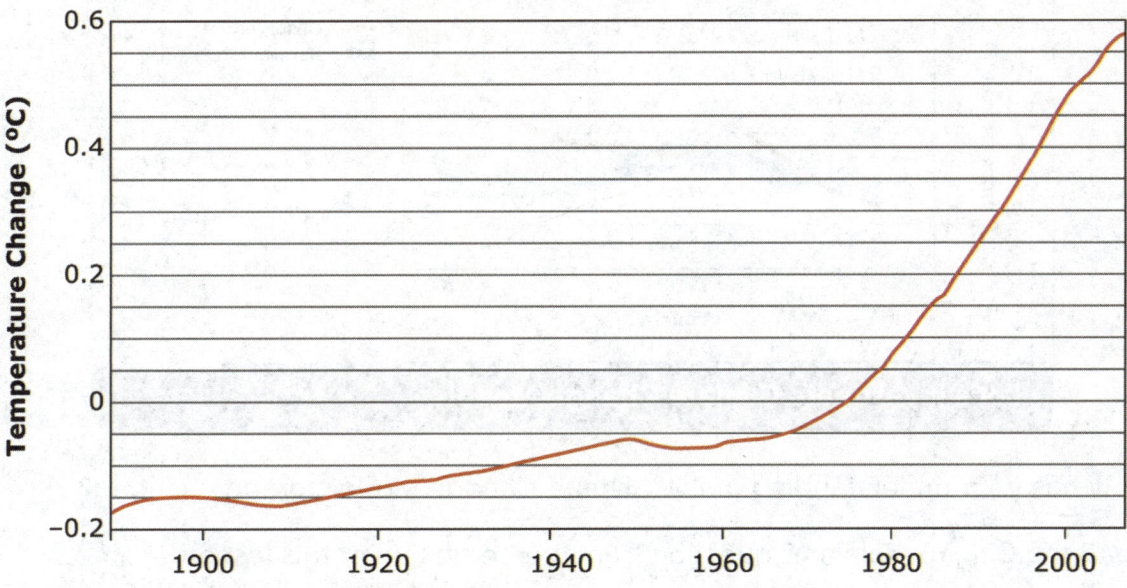

The graph shows the change in temperature from the year [1800/1850/1890/1900] to the year [1995/2000/2005/2010]. The overall change is about [0.2°C/ 0.6°C/0.8°C]. From 1900 to 1960, the temperature changed about [0.1°C/ 0.2°C /0.5°C/ 1°C]. A more rapid increase began in about [1950/1970/2000/2005].

SEP Analyzing and Interpreting Data

CONCEPT 4.1 | Why don't Earth's natural processes account for the observed global climate change? How can we determine what is causing the change?

Activity 19
Record Evidence
Climate Change Models

Quick Code
ca6775s

As you worked through this lesson, you investigated and gathered evidence about causes of climate change. Now, take another look at the Climate Change Models graph, which you first saw in Engage.

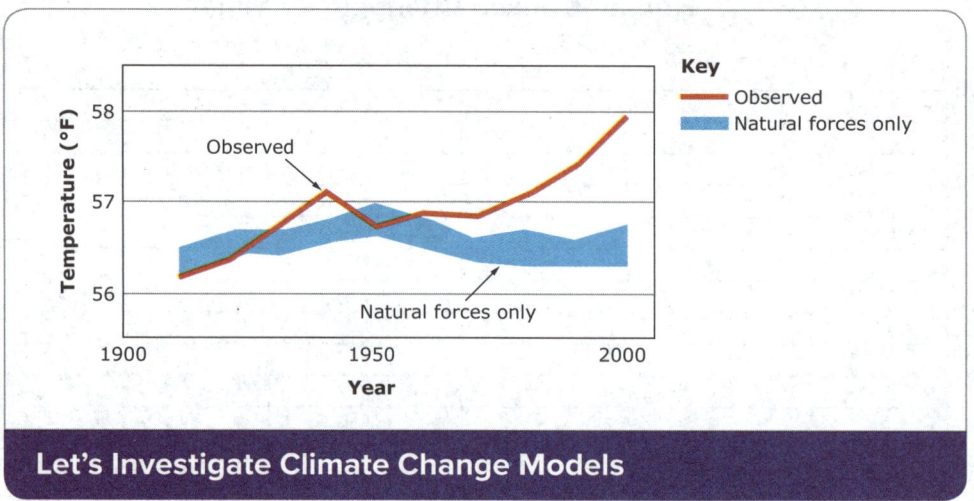

Let's Investigate Climate Change Models

How has your understanding of the Climate Change Models graph changed?

Read the Can You Explain? question from the beginning of this lesson.

 Can You Explain?

Why don't Earth's natural processes account for the observed global climate change? How can we determine what is causing the change?

Plan a scientific explanation to answer the Can You Explain? question. To help you formulate your scientific explanation, fill in the graphic organizer to organize your ideas. Recall that a scientific explanation contains three elements: a scientific claim, evidence to support the claim, and reasoning that connects the evidence to the claim. Think about and decide how you can use models or other representations as evidence to support your claim. These can include graphs, charts, drawings, photographs, or other types of graphics. Include these graphics in the Evidence section of the table.

Scientific Claim

Evidence to Support Claim

Reasoning That Connects Evidence to Claim

SEP Constructing Explanations and Designing Solutions

CONCEPT 4.1 | Why don't Earth's natural processes account for the observed global climate change? How can we determine what is causing the change?

Write your scientific explanation in the spaces provided. Be sure to include at least one graphic as part of your evidence. You can either draw it or print it out and attach it.

STEM in Action

Activity 20
Analyze

Climate

Quick Code
ca6776s

Write what you know and what you would like to know about prehistoric climate change in the chart below. Continue to fill out the chart as your read the text and after you have finished the text.

What I Know	What I Need to Know	How Can I Find Out?

SEP Obtaining, Evaluating, and Communicating Information

Climate

Have you ever wondered what Earth's climate was like billions of years ago? How much it has changed? What could cause climate to change before people were around? These are the types of questions that planetary scientists try to answer. James Pollack, a gay Princeton- and Harvard-educated astrophysicist at NASA, was a pioneer in this field. He was a student of Carl Sagan and worked with his mentor throughout the 1960s to develop and refine models of the atmospheres of Venus and Mars. They made predictions based on these models, which were later verified by the Mariner and Viking spacecraft.

Over his career, Pollack used models of other planets to help us understand how Earth's climate developed. You have probably heard how an enormous asteroid struck Earth and caused the extinction of the dinosaurs. Pollack figured out how the debris from that impact insulated Earth and created a "nuclear winter." From this work, we have learned how nuclear weapons can trigger similar effects.

Volcanic Eruption

ELABORATE

Concept 4.1: Causes of Climate Change

Climate *cont'd*

Pollack also made important contributions to our understanding of how volcanic eruptions can affect Earth's climate. Volcanic eruptions eject material high into the atmosphere. This material includes gases such as carbon dioxide, sulfur, and methane, which can cause severe global warming. Earth has had many periods of high volcanic activity that affected the atmosphere. For example, over 200 million years ago, intense volcanic activity led to the extinction of 76 percent of all marine and terrestrial species. This extinction is actually what allowed the dinosaurs to thrive!

Pollack was a brilliant scientist who won many awards and honors. One of these was the prestigious Gerard P. Kuiper Prize of the Division for Planetary Sciences of the American Astronomical Society. He was known for his brilliance, creativity, and enthusiasm. You can learn more about him in the book *Carl Sagan: A Life* by Keay Davidson.

Prehistoric Climate Change

Highlight the correct term to complete each sentence below.

Climate changes can be caused by natural forces such as volcanoes, which can _____ [warm/cool] the planet by ejecting _____ [gases/dust] high into the atmosphere. Large asteroid impacts can _____ [warm/cool] the planet by ejecting _____ [gases/dust] into the atmosphere.

CONCEPT 4.1

Why don't Earth's natural processes account for the observed global climate change? How can we determine what is causing the change?

Activity 21
Concept Review

Review: Causes of Climate Change

Quick Code
ca6778s

Now that you have completed the objectives for this concept, review the core ideas you have learned. Record some of the core ideas below.

Core Ideas

_____

_____

_____

_____

_____

_____

_____

_____

_____

_____

_____

Talk with a Group

Now, think about the Measuring Climate Change map you saw in Get Started. Discuss how what you've learned about causes of climate change can help you understand the Measuring Climate Change map.

Concept 4.1: Causes of Climate Change

CONCEPT
4.2

Climate Change Impacts Organisms

Student Objectives

By the end of this lesson:

☐ I can develop a model to describe and predict the effects of habitat destruction on a population of living things.

☐ I can argue from evidence that habitat destruction occurs and negatively impacts the dynamic equilibrium that exists within and between populations of living things.

☐ I can analyze data to describe and predict how climate change affects the stability of habitats and how living things respond to environmental stimuli.

Key Vocabulary

behavior, deforestation, environment, global warming, habitat, hibernate, migration, predator, prey, stimulus

Quick Code
ca6780s

Activity 1
Can You Explain?

How might climate change impact the behavior of the monarch butterfly?

Quick Code
ca6781s

_____
_____
_____
_____
_____
_____
_____
_____
_____
_____
_____

CONCEPT 4.2
How might climate change impact the behavior of the monarch butterfly?

Activity 2
Ask Questions

Migration Journey of the Monarch Butterfly

Monarch butterflies migrate yearly from Canada down to Mexico. Monarchs west of the Rocky Mountains may overwinter in southern California. In this concept, you will consider how the environment plays a role in animal behaviors such as migration and hibernation.

Watch the video Migration Journey of the Monarch Butterfly. Then, **answer** the questions that follow.

Quick Code
ca6782s

Let's Investigate Migration Journey of the Monarch Butterfly

SEP Asking Questions and Defining Problems
CCC Cause and Effect

What factors are needed for a monarch butterfly to survive? What features make Mexico a good winter habitat for butterflies?

___

Why do monarch butterflies migrate?

___

What questions do you have about climate and migration?

___

Concept 4.2: Climate Change Impacts Organisms

CONCEPT 4.2 | How might climate change impact the behavior of the monarch butterfly?

Activity 3
Evaluate

What Do You Already Know About How Climate Change Impacts Organisms?

Species Connections

Imagine the following habitat. A grassland is being plowed to create a local park. The species that depend directly on the grass for food and shelter are obviously affected. Explain why other species are also affected. **Write** your response below.

Quick Code
ca6783s

_____
_____
_____
_____
_____
_____
_____
_____
_____
_____
_____
_____

Causes of Destruction

In your opinion, what are the top five causes of habitat destruction? **Create a list** that ranks the top five causes of habitat destruction. Number one on your list should be the cause that leads to the greatest destruction.

CONCEPT 4.2 How might climate change impact the behavior of the monarch butterfly?

Explain Your Answer

Now, **explain** the thinking behind your answer. Why did you select the causes you did, and why did you place them in the order you did?

Migration

Which statements about migration are true? **Place a check** next to all that apply.

☐ Organisms only migrate during the winter season.

☐ Only organisms whose parents migrated can migrate.

☐ Organisms can migrate due to habitat destruction.

☐ Large land animals always migrate when predators are present.

☐ Organisms can migrate due to a natural disaster.

CONCEPT 4.2 | How might climate change impact the behavior of the monarch butterfly?

How Might Climate Change Affect Habitats?

Activity 4
Observe

Introduction to Habitats

Watch the video Introduction to Habitats and **answer** the questions that follow.

Quick Code
ca6786s

Introduction to Habitats

CCC Stability and Change

What can you assume about animals that live in the same habitat?

_____
_____
_____
_____
_____
_____
_____

What might happen if one species is removed from a habitat?

_____
_____
_____
_____
_____
_____
_____

CONCEPT 4.2 | How might climate change impact the behavior of the monarch butterfly?

Activity 5
Design Solutions

Hands-On Engineering: Habitat Destruction

Quick Code
ca6787s

In this investigation, you will design and build a model to demonstrate how habitat destruction affects organisms. You will develop a plan, construct a model, and use the model to simulate the effects of destroying a habitat. Then, you will evaluate the effectiveness and limitations of your model.

Predict

How can we create a model that demonstrates the effect of habitat destruction on local organisms?

_____

_____

_____

_____

_____

_____

_____

_____

_____

_____

_____

What materials do you need? (per group)

- Shoeboxes
- Cardboard boxes
- Large bin, with lid
- Modeling clay
- Toothpicks
- Craft sticks
- Tissue paper
- Construction paper
- Markers
- Colored pencils
- Scissors
- String

Procedure

In groups of three, discuss what types of things would help show habitat destruction in a scaled-down model, either by sketching your ideas on a piece of paper or writing down some of the details that you want to be sure to include in your model. The model should involve a collection of living and nonliving things found in your local environment that can be manipulated to show the effect of habitat destruction. Develop and build your model.

CONCEPT 4.2

How might climate change impact the behavior of the monarch butterfly?

Reflect

What type of habitat does your model represent? What types of organisms live in that habitat?

_____
_____
_____
_____
_____
_____
_____

What are some events or human activities that might damage or destroy the habitat in your model?

_____
_____
_____
_____
_____
_____
_____

How would organisms be affected by the destruction of the habitat your model represents?

_____
_____
_____
_____
_____

In what ways does your model accurately demonstrate habitat destruction and its effects? What are some limitations of your model?

_____
_____
_____
_____
_____
_____
_____
_____
_____
_____

Concept 4.2: How might climate change impact the behavior of the monarch butterfly?

Experiment with "destroying" the habitats you built in the previous activity. Present the finished models to the rest of the class, incorporating some of the ideas learned in this lesson. Discuss the benefits and limitations of your model below.

How was your model like the real phenomenon of habitat destruction, and how was it lacking?

_____
_____
_____
_____
_____
_____
_____
_____

Is there anything you could do differently to improve your model?

_____
_____
_____
_____
_____
_____
_____

In general, what are the benefits and limitations of using scientific models?

CONCEPT 4.2 | How might climate change impact the behavior of the monarch butterfly?

Activity 6
Observe

Got Habitat?

Quick Code
ca6788s

Use the online exploration Got Habitat? to learn about how animals, such as the cougar population in the Santa Ana Mountains, have habitat preferences that determine how they will survive the effects of habitat loss.

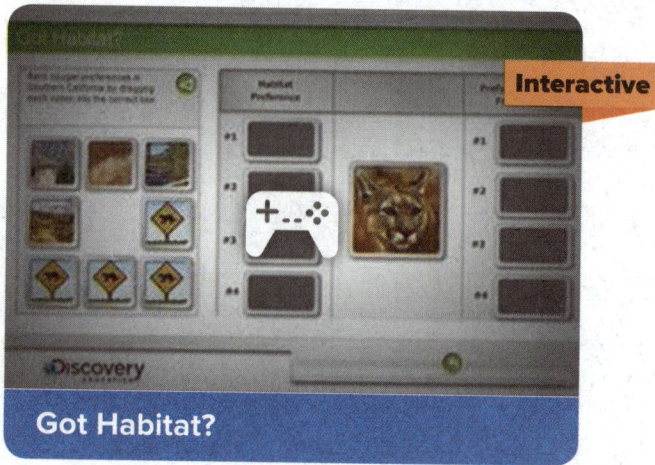

Got Habitat?

Go Online to explore the habitat of the cougar. Then, **answer** the questions that follow.

List several reasons why wildlife habitat is shrinking.

_____

_____

_____

_____

_____

Why do cougars prefer riparian habitat?

What is the meaning of *corridors* as used in this exploration?

After exploring the results of this scientific study, what would you recommend to save the cougar's habitat?

Activity 7
Analyze

Climate and Habitat Destruction

Quick Code
ca6789s

As you read the text, **highlight** examples of natural processes or factors that can lead to habitat destruction and underline examples of human activities that can lead to habitat destruction. Then, **record** the most important evidence in the T-Chart.

Review the Design Solutions activity Habitat Destruction. Add any additional important factors that can lead to habitat destruction from this activity to your chart, if they are not already there.

Natural Processes	Human Activities

Climate and Habitat Destruction

A **habitat** is a place where an organism lives. Habitats can be deserts, forests, streams, oceans, and more. Despite the variety, all habitats provide four things for an organism: food, water, shelter, and space. When one of those resources is depleted or taken away entirely, it is called habitat destruction. But why is it called *habitat destruction* if only one resource is taken away? It turns out these resources often depend on one another. It is almost impossible to make a single change to a habitat and expect a single reaction. In this concept, you will explore some of the consequences of habitat destruction.

Earth has a dynamic **environment**, which means that it is constantly going through changes. Many of these natural changes can cause habitat destruction. For example, hurricanes, fires, floods, volcanic eruptions, and earthquakes can all be destructive. Diseases can also wipe out populations of plants and animals, which can destroy the shelter and the availability of food for many organisms.

When populations overpopulate—producing more offspring than can be sustained in the habitat—habitat destruction can occur. Overpopulation leads to less food, water, shelter, and space for other populations that live in the area. When new species come into an area, whether naturally or brought by humans, they can become invasive species. Invasive species use up the resources that native populations need and can even become the dominant population.

Although many forms of habitat destruction are natural, human activities can cause or accelerate habitat destruction.

Climate and Habitat Destruction *cont'd*

You have learned how many of the activities humans engage in contribute to global warming and climate change. These activities seem to be good for the human population, but they cause or accelerate habitat destruction.

Because humans live in habitats, we should be very careful in our choices. For example, agriculture can destroy native habitats to make new grazing land or new fields for crops. Large farms can also lead to nutrient runoff. This does not sound like such a bad thing until you look at its consequences. Extra nutrients in the soil are often washed into nearby waters, and algae thrive off those extra nutrients, causing algae overpopulation. This causes many plants and aquatic lifeforms to die off. Increased severe weather events, like hurricanes and floods caused by climate change, exacerbate runoff issues.

Development can also be harmful to habitats when hills, prairies, and valleys are turned into factories, homes, and roads. Human pollution and waste disposal have led to landfills where habitats once were, adding greenhouse gases to the atmosphere and ultimately leading to melting ice caps and poisonous water and air. Industry has led to **deforestation** and lands ripped up for mining. On top of that, humans have introduced many invasive species, killing off native plants and animals.

Plants and animals depend on their habitats for space, food, shelter, and water. Humans are increasing the rate at which the climate is changing on Earth, which is changing the habitats upon which we all depend. When habitats change, plant and animal populations respond by changing their **behavior** to adapt to the new habitat. If populations are not able to adapt or move, they risk becoming extinct.

Activity 8
Evaluate

Habitat Destruction

Quick Code
ca6790s

Choose which student's statement most accurately describes habitat destruction. **Circle** the student's name whose statement is most accurate.

Chuck: Habitat destruction happens, but only because climate change is increasing the number of tornadoes that destroy land habitats.

Justine: Habitat destruction can only happen when human-made events, like construction, occur.

Marlene: Habitat destruction can happen to lakes and oceans, but it rarely affects land habitats.

Serena: Habitat destruction can happen because of climate change, human-made events, and natural events.

Trevor: Habitat destruction happens because of natural events, like hurricanes.

List the evidence that supports the correct statement and the evidence that shows why the other statements are incorrect.

_____
_____
_____
_____
_____
_____
_____

CCC Cause and Effect

How Does Climate Change Affect Organism Behavior?

Activity 9
Analyze

Climate and Organism Behavior

Quick Code
ca6791s

Before you read the text that follows, **brainstorm** some answers to the questions below. After you have read the text, you may want to look back and revise your answers.

How does a rooster know to crow in the morning? How does a wolf know to howl at the moon? How does a goose know where to migrate in the winter?

_____
_____
_____
_____
_____
_____
_____

Read the following text and **highlight** or **underline** any information you already knew from previous concepts.

Climate and Organism Behavior

All organisms exhibit behavior. Behavior is the pattern by which organisms respond to **stimuli** from their environment. Behavior can be the way a **predator** responds to the scent of **prey**, the way a person responds to music, or the way a worker bee responds to the waggle dance.

Instinctive responses are automatic, natural responses to the environment. Many instinctive responses are tied to the basic needs of organisms, like safety, food, shelter, or water. The **migration** of birds is an instinctive response. It helps them survive by moving to warmer climates. Monarch butterflies migrate from the northern **latitudes** in the United States to southern latitudes in Mexico each year in the fall. Many animals migrate, including species of mammals such as whales and caribou, as well as some birds and some species of fish.

Hibernation is an instinctive response that allows organisms to adapt to colder weather when food is scarce. As winter approaches, animals such as bears, hedgehogs, frogs, and chipmunks experience a slowdown in their metabolisms. When their metabolisms slow, they require less food for their daily functions and can conserve their energy. Without this behavior, they would not be able to satisfy their energy needs with available resources and would not survive the winter.

CONCEPT 4.2 | How might climate change impact the behavior of the monarch butterfly?

Activity 10
Observe

Migration and Hibernation

Quick Code
ca6792s

Watch the video. Then, **answer** the questions that follow.

Come up with three definitions for *migration* and *hibernation*.

_____
_____
_____
_____
_____
_____
_____
_____

Write a brief explanation of why animals migrate or hibernate.

CONCEPT 4.2 — How might climate change impact the behavior of the monarch butterfly?

Activity 11
Interpret Data

Grasshopper Habitats Shift in Latitude

Quick Code
ca6793s

Many instinctive responses revolve around the climate. So how does an organism's behavior change when the stimuli from the environment change? Consider a grasshopper. It is a cold-blooded insect, meaning it cannot regulate its own body temperature. It takes on the temperature from its surroundings. How would you expect a population of grasshoppers to react as the average temperature increases?

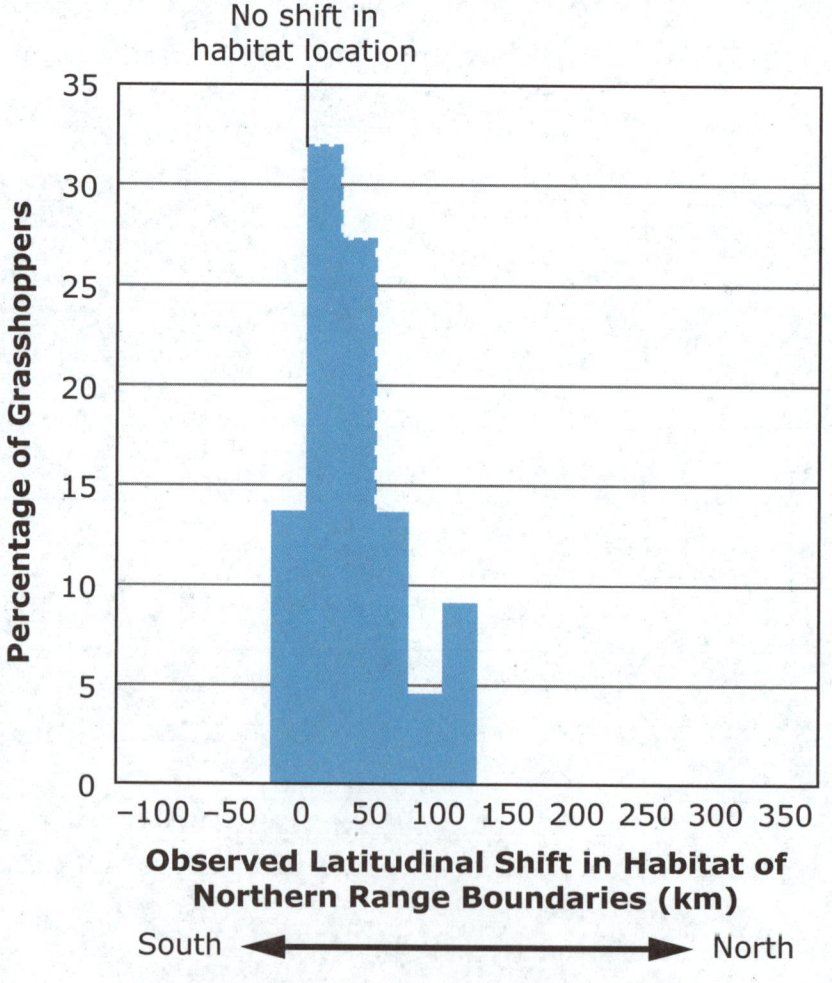

SEP Using Mathematics and Computational Thinking

CCC Stability and Change

Photo credit: GomezDavid / E+ / Getty Images

© Discovery Education | www.discoveryeducation.com

What conclusions can you draw from this representation of grasshopper habitats?

Activity 12
Analyze

Shifting Habits and Global Warming

Quick Code
ca6794s

As you read the following text, **complete** a T-Chart that shows how global warming and climate change are affecting animal migration and hibernation.

Shifting Habitats and Global Warming

A changing climate can cause populations to shift their habitats. Recall how climate regions get cooler as you move farther away from the equator. As human activities continue to warm our planet, animals like the grasshopper take advantage of these differences and may shift their locations to reach more favorable temperatures.

Climate change is causing polar bears, seals, and walruses to swim farther to find meals. Warming trends are causing ice floes to melt earlier. The distance between floating sea ice is greater. Because these animals use this ice as launching pads for hunting, they must swim farther to get food. This requires more energy. Polar bears, seals, and walruses must hunt more to build their fat storage. Lower-weight animals have a more difficult time feeding their young and surviving in cold temperatures. As a result, this change in climate threatens the survival of the species.

Global warming is changing migratory animals' behavior. For example, some migratory birds are beginning their return trips back north earlier as temperatures increase earlier in the year.

Those traveling great distances risk returning to a habitat that is still too cold or that lacks food sources needed for survival.

Temperature changes are affecting breeding behaviors as well. Scientists have observed that light-bellied brent geese are laying more eggs when the temperatures are warmer. Females are staying with their eggs longer, resulting in more offspring being born. This comes at a cost, however. More female light-bellied brent geese are dying during these warmer years. Females are spending more energy to lay more eggs. And because the birds lay their nests on the ground, the females are more easily hunted by predators.

Hibernating animals' behavior is also affected by climate change. During hibernation, an animal's body slows down to conserve energy. It survives off its fat stores throughout the winter. The colder the temperature, the slower the body burns energy. When winter temperatures are warmer, hibernating animals will burn through their fat stores sooner. Animals like bears and snakes begin to awaken from hibernation earlier. When that happens, they may find that there is too little food to sustain them.

Because everything is connected in Earth's system, the precise consequences of global warming can be difficult to model. However, scientists know that the effects are far-reaching and difficult to reverse. It is true that some ecosystems and some organisms are more affected by climate change than others. Yet, climate change is a global problem. With international effort, we can slow the progress of global warming and address its long-term impacts.

Animal Migration	Hibernation

Activity 13
Evaluate

Changing Climate Changes Behaviors

Quick Code
ca6795s

Number the statements in the correct sequential order (1–6) to show how increasing temperatures can affect walruses in the Arctic.

_____ Walruses lose weight.

_____ Walruses burn more energy.

_____ More walruses and their pups die.

_____ Walruses travel farther to find food.

_____ Walruses become endangered or extinct.

_____ Sea ice melts, leaving smaller patches of ice spaced farther apart.

CCC Stability and Change

CONCEPT 4.2 | How might climate change impact the behavior of the monarch butterfly?

Activity 14
Record Evidence

Migration Journey of the Monarch Butterfly

Quick Code
ca6796s

As you worked through this lesson, you investigated and gathered evidence about how climate change impacts organisms. Now, take another look at the video Migration: Journey of the Monarch Butterfly, which you first saw in Engage.

Let's Investigate Migration: Journey of the Monarch Butterfly

How has your understanding of the Migration Journey of the Monarch Butterfly video changed?

Read the Can You Explain? question from the beginning of this lesson.

> **Can You Explain?**
>
> How might climate change impact the behavior of the monarch butterfly?

Use your new understanding of the video Migration Journey of the Monarch Butterfly to write a scientific explanation answering a question.

1. **Choose** a question. You may choose to answer the Can You Explain? question or one of the questions you wrote at the start of this lesson.

2. Use the following information to help you write your explanation.

With a partner, plan a scientific explanation by writing a claim, listing evidence, and explaining your reasoning in the graphic organizer.

Scientific Claim

Evidence to Support Claim

Reasoning That Connects Evidence to Claim

SEP Constructing Explanations and Designing Solutions

Concept 4.2: Climate Change Impacts Organisms

CONCEPT 4.2 — How might climate change impact the behavior of the monarch butterfly?

Write your scientific explanation in the spaces provided.

 in Action

Activity 15
Analyze

Migrating Sand Dunes

Quick Code
ca6797s

Read the following text, **watch** the video, and **highlight** the factors that impact the rate of dune migration.

Migrating Sand Dunes

Have you ever been to a desert? If so, you may have seen fields of sand dunes far into the distance. They may look very quiet and peaceful, but sand dunes are actually very active. Sand is constantly being picked up and carried by the wind, and then deposited farther away. This causes the dunes to migrate, or move, in the direction that the wind is blowing. What would happen if your home was in the way? This is exactly what is happening on the Navajo Indian Reservation in the southwestern United States.

The Navajo people of the southwestern United States are strongly rooted to the land. They

A Sand Dune Field

grow fruits and vegetables and raise sheep, cattle, and other animals. Migrating sand dunes are now threatening their way of life. Dr. Margaret Hiza Redsteer is a Native American geologist who is studying these dunes. By monitoring precipitation, wind, and vegetation, she has shown that climate change is causing dune migration. Precipitation has decreased significantly in recent decades. The soil has become too dry to support the plants that stabilize the dunes. Bare, dry sand is more easily eroded than sand covered by plants. Wind has also increased. Stronger winds move more sand farther and faster.

By comparing aerial photographs that were taken over the past 50 years, Dr. Redsteer estimates that dunes in the area have moved at rates as high as 112 to 157 feet per year (34 to 48 meters per year). Dunes have already covered roads and rangelands and are now threatening homes and villages. Some villages will probably need to be abandoned. The people will need to move elsewhere. The results of Dr. Redsteer's work will help the Navajo people understand dune migration and identify ways to mitigate its negative effects.

Dune Research

ELABORATE

Concept 4.2: Climate Change Impacts Organisms | 311

Dune Research

Think about what you have read and watched. Use what you have learned to **answer** the question that follows.

How are people on the reservation affected by dune migration?

_____

_____

_____

_____

_____

_____

Effects of Climate Change

In the southwestern United States, precipitation has been decreasing and temperatures have been increasing over the past 50 years. These weather trends are causing some things to increase and other things to decrease. Write whether each item in the list below will increase or decrease. Be prepared to discuss your reasoning.

- Extent of rangeland
- Agricultural production
- Lake levels
- Soil moisture
- Water resources
- Dune migration rate
- Size of dune fields

Increase	Decrease

Concept 4.2: Climate Change Impacts Organisms

CONCEPT 4.2
How might climate change impact the behavior of the monarch butterfly?

Activity 16
Concept Review

Review: Climate Change Impacts Organisms

Quick Code
ca6800s

> Now that you have completed the objectives for this concept, review the core ideas you have learned. Record some of the core ideas below.

Core ideas

_____
_____
_____
_____
_____
_____
_____
_____
_____
_____
_____
_____
_____

Talk with a Group

Now, think about the Measuring Climate Change map you saw in Get Started. Discuss how what you've learned about how climate change impacts organisms can help you understand the Measuring Climate Change map.

EVALUATE

Concept 4.2: Climate Change Impacts Organisms

CONCEPT 4.3

Reducing Human Impacts on the Environment

Student Objectives

By the end of this lesson:

- [ ] I can develop a model to describe and predict how pollution affects groundwater and surface water locally and globally.

- [ ] I can analyze and interpret data to describe various causes of drinking water pollution.

- [ ] I can develop a model to describe and predict the effect of land exploitation on biodiversity locally and globally.

- [ ] I can obtain information about strategies humans can use to maintain biodiversity within their local ecosystems and minimize global changes on Earth, evaluate the benefits and obstacles of each strategy, and communicate the most effective strategies.

Key Vocabulary

biodiversity, deforestation, ecosystem, environment, pollution, remote sensing

Quick Code
ca6802s

Concept 4.3: Reducing Human Impacts on the Environment

Activity 1
Can You Explain?

How can you reduce the impact cattle have on climate change and the environment?

Quick Code
ca6803s

CONCEPT 4.3

How can you reduce the impact cattle have on climate change and the environment?

Activity 2
Ask Questions

Cattle Population in the United States

Quick Code ca6804s

Review the following images, and then **answer** the questions that follow.

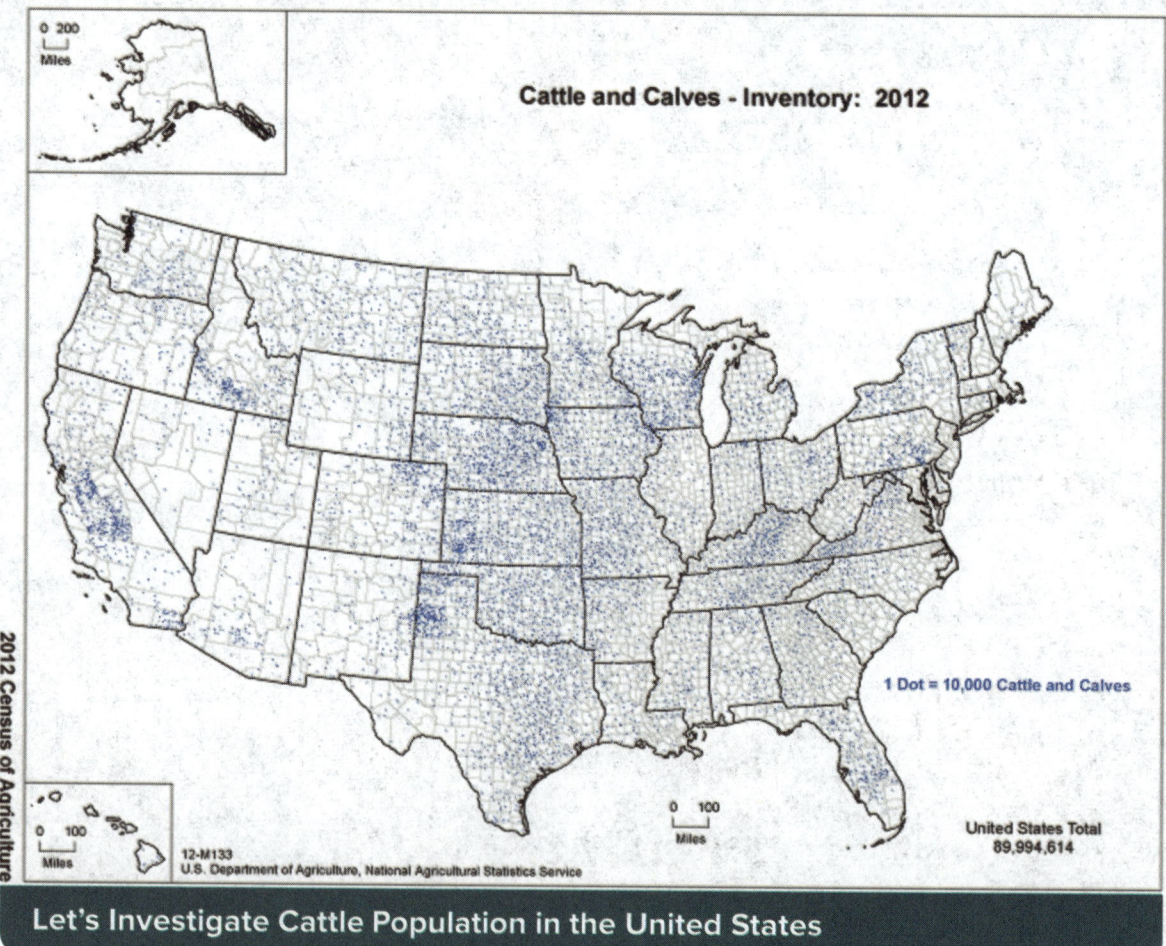

Let's Investigate Cattle Population in the United States

- **SEP** Asking Questions and Defining Problems
- **SEP** Analyzing and Interpreting Data
- **CCC** Patterns

Traffic

Vegetables at Market

Irrigation Sprinklers

Concept 4.3: Reducing Human Impacts on the Environment

CONCEPT 4.3

How can you reduce the impact cattle have on climate change and the environment?

On a separate sheet of paper, create four blocks: one for each image. Write one sentence in each block about each of the images. Then, write one sentence that incorporates all four images.

Your Ideas

Look again at the map of cattle populations in the United States. What kinds of impacts do you think raising so much cattle would have on the environment?

What other questions do you have about how cattle or agriculture affect the environment?

CONCEPT 4.3

How can you reduce the impact cattle have on climate change and the environment?

Activity 3
Evaluate

What Do You Already Know About Reducing Human Impacts on the Environment?

Quick Code
ca6805s

Environmental Problems

Identify each activity according to whether it results in habitat loss, pollution, or both. **Write** the correct category on the line next to the statement.

Building a dam to generate electricity _____

Emissions from cars and trucks _____

An oil spill that covers a large area of the ocean and shoreline
_____

Clearing a forest for housing _____

True or False?

Read the following statements. **Choose** all of the statements that are true.

Too much damage has been done to the environment for us to make a difference today.

Small efforts, such as recycling programs, can add up and have an impact.

All a company needs to do to protect the environment is follow current environmental laws.

Pollution on the Move

Pollution travels from one resource to another. **Answer** the questions by describing one example for each type of pollution movement.

How can pollution travel from air to water?

_____

_____

_____

How can pollution travel from water to air?

_____

_____

_____

How can pollution travel from land to water?

_____

_____

_____

CONCEPT 4.3 | How can you reduce the impact cattle have on climate change and the environment?

How Do Scientists Monitor Populations for Signs of Pollution, and Why Is This Important?

Activity 4
Reason

Causes and Effects of Water Pollution

In this activity, you will develop a model to illustrate how human activity affects water quality.

Quick Code
ca6808s

What materials do you need? (per group)

- Water
- Plastic container, 2 qt
- Sand
- Feathers
- Aquarium pebbles
- Vegetable oil
- Chenille sticks
- Craft sticks
- Modeling clay
- Sponges
- Plastic straws
- Twigs
- Flowers
- Cotton balls

SEP Developing and Using Models

Procedure

Part 1: Pollution Videos

As you **watch** each video, **complete** the chart that follows. **Look** for evidence that shows how human activity can cause water pollution and how water pollution affects the environment and people.

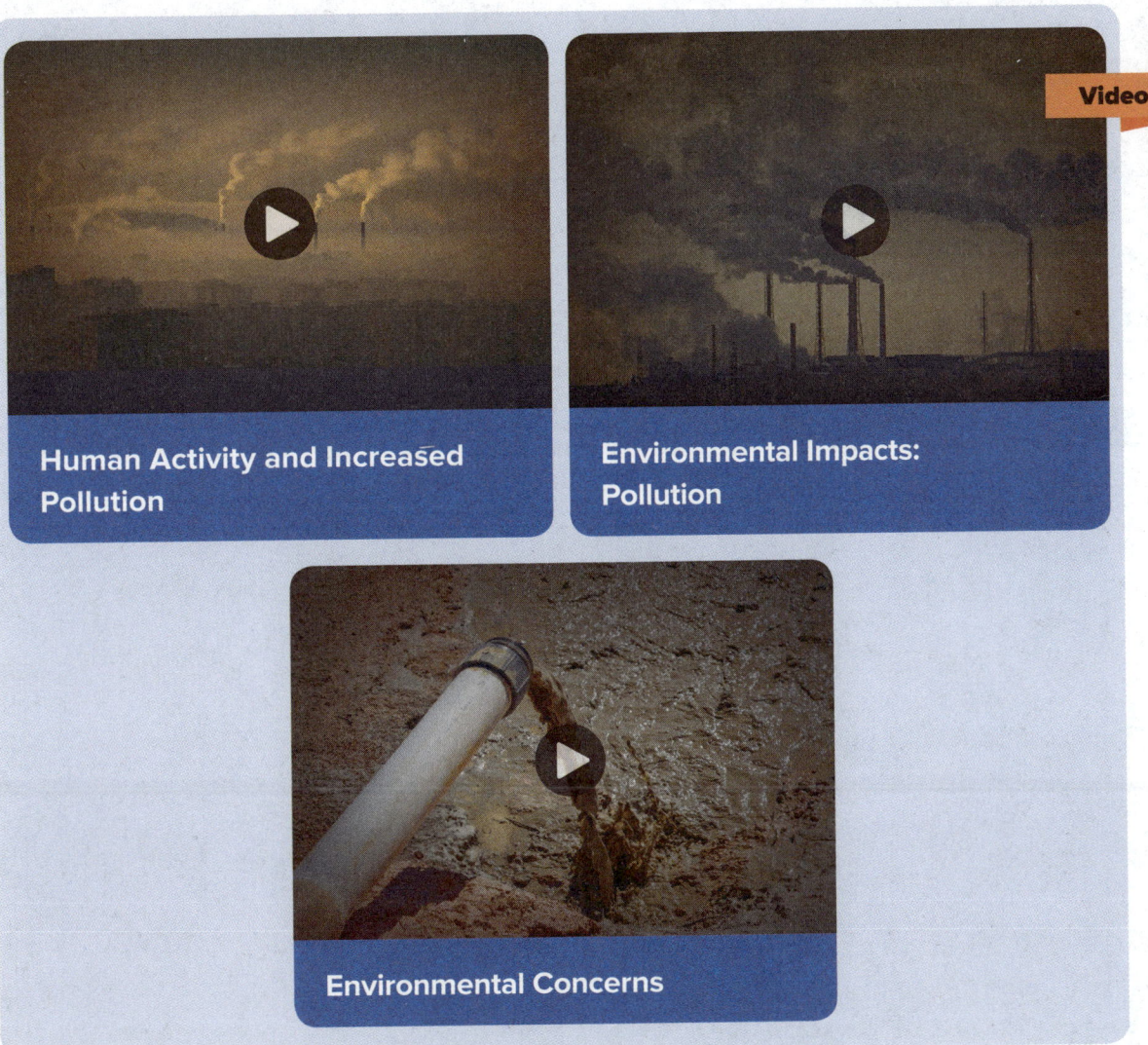

CONCEPT 4.3 | How can you reduce the impact cattle have on climate change and the environment?

Complete the chart based on the videos. You will use this information to create your model.

Causes of Water Pollution	Effects of Water Pollution

Brainstorm what specific human activity (cause) and consequence (effect) you want to model. Determine how you will construct a model with the materials available. **Sketch your design, and then begin building your model.**

Reflect

How does building a model that demonstrates the causes and effects of pollution, rather than reading about it or looking at pictures, help you understand how pollution occurs?

_____
_____
_____
_____
_____

CONCEPT 4.3 | How can you reduce the impact cattle have on climate change and the environment?

What new information did building your model teach you about the effects of human activities on water pollution?

_____

_____

_____

_____

Explain how the effects of the type of pollution you modeled could be minimized.

_____

_____

_____

_____

_____

Activity 5
Analyze

Monitoring Pollution

Quick Code
ca6809s

Read the following text. As you read, **highlight** information about sources of air pollution, water pollution, and waste.

Monitoring Pollution

Pollution is a harmful substance in the **environment**. It may also be poisonous. Pollution comes from many human activities, including energy use, farming, industry, and daily activities at home and at school. Pollutants in the air can make the air dangerous to breathe. They cause heating of the atmosphere as well. Pollutants in the ground can harm the soil. This hurts the plants and animals living there. Pollutants can also leak into nearby water. Water pollution harms many types of aquatic life. It can cause excess algae growth. It can also contaminate drinking water.

Pollution can be cleaned up with a lot of time and effort. Preventing pollution is more effective and efficient. Many governments have programs to reduce or eliminate pollution. For example, the Clean Water Act in the United States helps ensure that everyone has access to safe drinking water.

Water, Pollution, and Waste

Monitoring Pollution *cont'd*

However, the laws do not always work as intended. People should therefore stay informed about the effects of pollution and work together to keep their environment safe.

Scientists can assess the water, air, and soil quality for signs of pollution. This helps them monitor the health of the **ecosystem**. Scientists use sensors to learn what chemicals are in the air. They also tell how much of each chemical is present. Some of these sensors are small enough to be attached to a bicycle. Scientists also collect samples of water. These samples are tested for chemicals, metals, and other contaminants. Scientists use this information to tell people whether their water is safe to drink. They also test soil samples. They do this to find out if the land has been contaminated with farm waste, sewage sludge, or water pollution that has leached into the soil.

Monitoring ecosystems leads to greater understanding of how much pollution an ecosystem can handle before living things are harmed. This information can help support laws that affect how much pollution a human population can produce.

Complete the 3-Way Venn diagram comparing air pollution, water pollution, and waste disposal based on the information you gathered from the text and video.

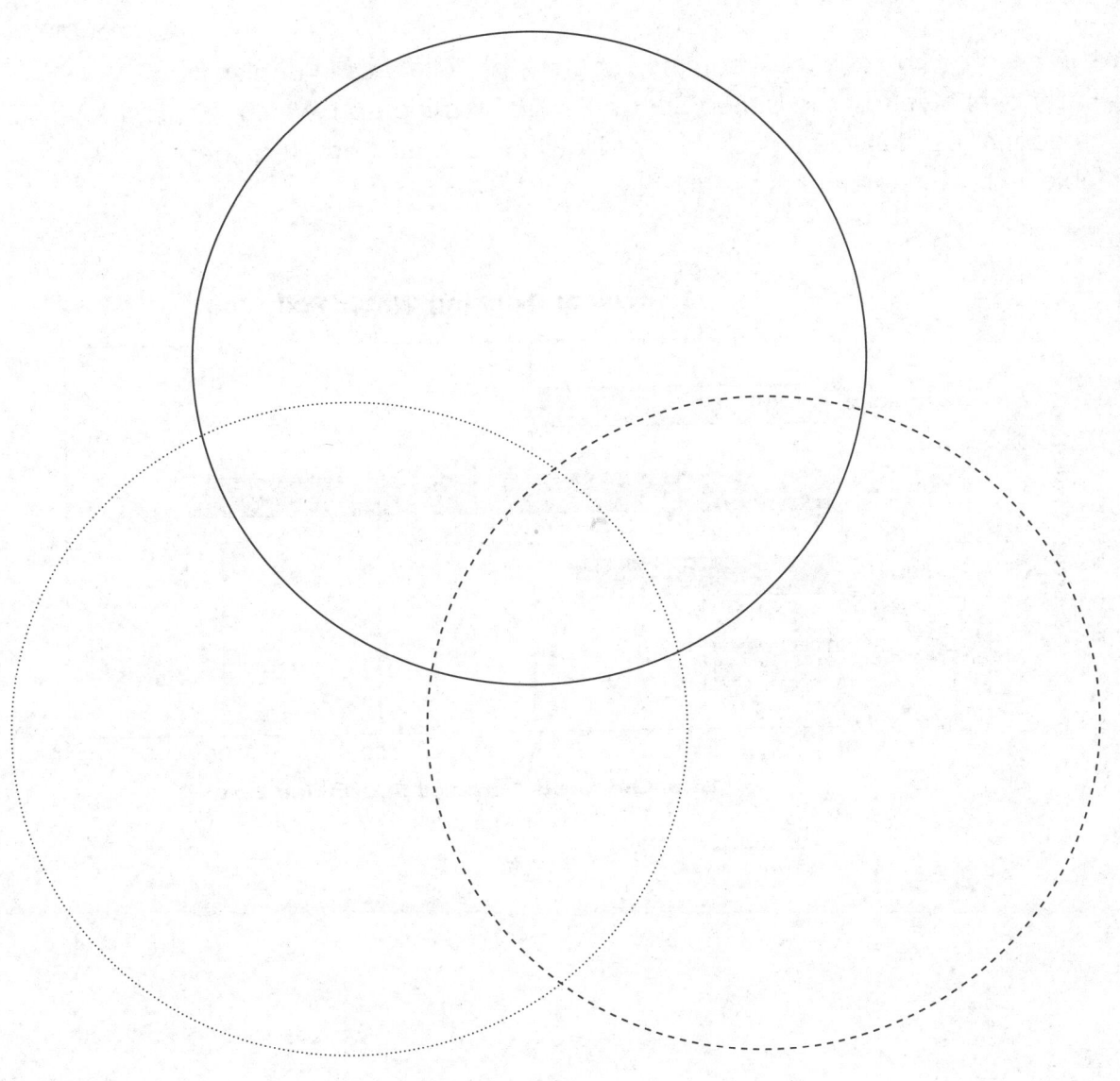

Concept 4.3: Reducing Human Impacts on the Environment | 333

CONCEPT 4.3 | How can you reduce the impact cattle have on climate change and the environment?

Activity 6
Evaluate

Safe to Drink

Quick Code
ca6810s

This graph shows the most common sources of pollutants in drinking water over a year in the United States. Based on the graph, which category best describes each of the following actions? **Complete** the table with two statements in each row.

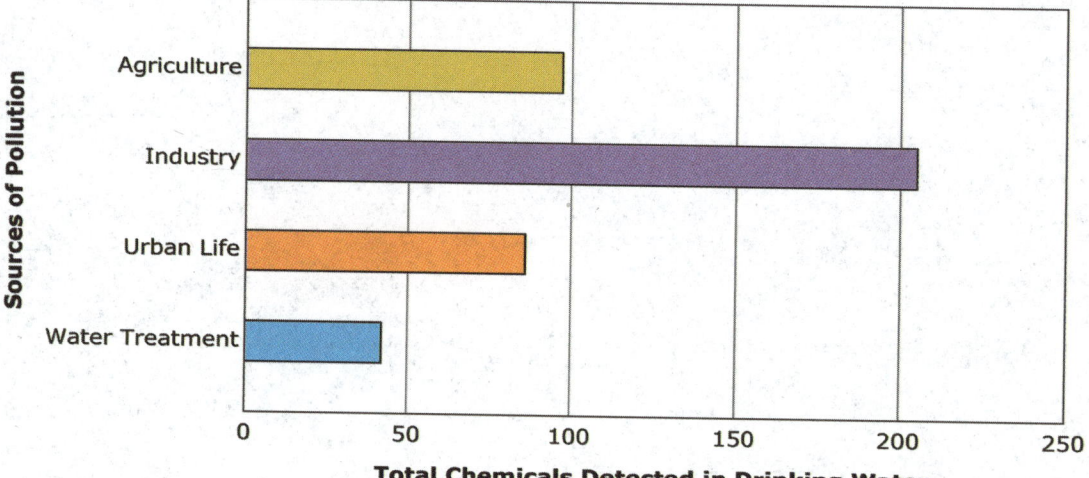

SEP Analyzing and Interpreting Data

- Animals on a farm produce solid waste.
- Crops are dusted with a chemical pesticide.
- A homeowner pours leftover paint down a storm drain.
- People litter as they walk through a park.
- Heavy rains cause a sewer system to flood.
- Smokestack emissions mix in the air to cause acid rain.
- A factory uses water to cool its machinery.
- Chemicals are added to drinking water to kill bacteria.

Category	Action 1	Action 2
Source with Most Pollution		
Source with 2nd Most Pollution		
Source with 3rd Most Pollution		
Source with Least Pollution		

How Can Preventative Measures Be Used to Reduce Land-Use and Environmental Problems?

Activity 7
Analyze

Land Use and Biodiversity

Quick Code
ca6811s

Read the following text. As you read, **think** about what *biodiversity* means and how it relates to ecosystem health.

Land Use and Biodiversity

Humans use land for many things. We use it to build homes and cities, we mine it for resources, and we use it to produce our food. Sometimes, an area is unsuitable for a needed use. In this case, people change the landscape instead of changing their plans. Examples include **deforestation** for crops or farm animals, draining swamps, and mountaintop mining. Although changing the land may make life easier for humans, habitat destruction can reduce **biodiversity** and have other unintended effects.

Biodiversity is the variety of life. This is one way to measure **ecosystem** health. All ecosystems have limited resources. Many of those are lost when land is developed. Unhealthy ecosystems have smaller populations and less biodiversity than healthy ones. What are the results of this decline? One result is the loss of plants and animals that could provide new medicines. Consider aspirin. It contains a painkilling chemical found in willow bark. Another possible result is that we have fewer kinds of foods. Many animal species pollinate food plants. They also kill plant pests. Can you think of other effects of biodiversity loss?

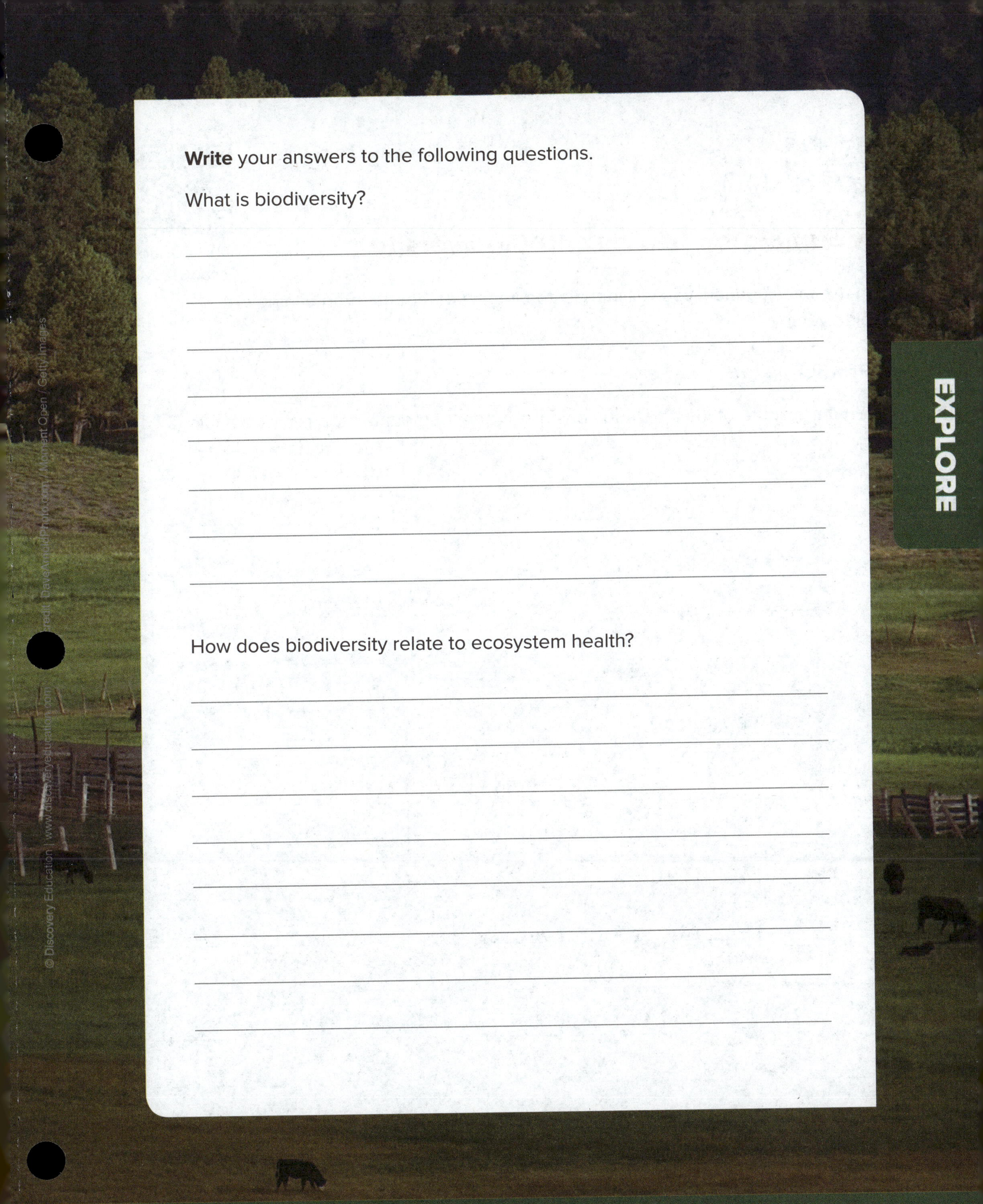

Write your answers to the following questions.

What is biodiversity?

_____
_____
_____
_____
_____
_____

How does biodiversity relate to ecosystem health?

_____
_____
_____
_____
_____
_____
_____

CONCEPT 4.3 | How can you reduce the impact cattle have on climate change and the environment?

Activity 8
Reason

Mountaintop Mining and Biodiversity

Quick Code
ca6812s

In this investigation, you will identify the impact of mountaintop mining on biodiversity.

Predict

What are some of the pros and cons of a model of mountaintop removal mining?

_____
_____
_____
_____
_____
_____
_____
_____
_____
_____
_____
_____
_____
_____

SEP Developing and Using Models

What materials do you need? (per group)

- Seeds, black beans
- Seeds, lima beans
- Seeds, pinto beans
- Bowls, plastic
- Paper plate
- Computer with Internet access
- Construction paper
- Forceps
- Large straws

Procedure

Part 1: Modeling Mountaintop Removal Mining

1. Use your hands to mix the three types of beans in the bowl until the beans appear evenly mixed.

2. Pour the beans onto a paper plate, forming a mound, which represents a mountain, and then cover it with the remaining lima beans.

3. Remove a handful of beans from the top of the "mountain." Sort the beans to see how many lima beans (living things) have been removed and how many black beans (coal) you collected. Repeat Step 2 if you did not collect any black beans.

Part 2: Comparing Ways to Get Coal

Design an investigation to test different ways to retrieve coal (black beans) from the mountain that will minimize the loss of biodiversity (lima beans).

Do a web search to investigate ways to mine coal. Try to find methods that prevent as much movement of the "mountain" as possible. Engineers and miners would also need to make such considerations in real life because some methods pose a risk to humans.

Concept 4.3: Reducing Human Impacts on the Environment

CONCEPT 4.3 How can you reduce the impact cattle have on climate change and the environment?

Reflect

How did the method of coal removal you explored affect biodiversity when compared to mountaintop removal?

_____
_____
_____
_____
_____
_____
_____

How were your models like the real thing? How were they different?

_____
_____
_____
_____
_____
_____
_____
_____

Based on what you observed, how do you think using renewable energy, such as solar or wind energy, compares to the impact of mountaintop mining on biodiversity?

_____

_____

_____

_____

_____

_____

_____

_____

Should people use mountaintop removal to retrieve coal? Why or why not?

_____

_____

_____

_____

_____

_____

_____

_____

Concept 4.3: Reducing Human Impacts on the Environment

Activity 9
Analyze

Land Use and Planning

Read the following passage and **underline** methods people use to prevent land-use problems.

Quick Code
ca6813s

Land Use and Planning

The United States is home to over 300 million people today. A population of 300 million people can put quite a bit of stress on the land and environment. If you read or watch the news, you might see stories about crowded cities, suburban sprawl, or the loss of animal habitats. All of these issues are related. We can't solve all of the problems that come with a large and growing population, but we can try to minimize our impact on the environment. Through careful planning and thoughtful use of the land, we can help prevent many of the problems affecting the world around us.

A large population can put stress on the environment. We must confront several issues related to how we currently use land. As a city grows, its boundaries are pushed outward, infringing on the land around the city. This sprawl can have serious consequences on the surrounding area. As vegetation is removed to make way for houses, the soil can be damaged, leading to more erosion and a greater likelihood of flooding. As the land use changes from rural to urban, native species can be forced out by construction. This can cause animals to die, or it can allow non-native (invasive) species to thrive.

SEP Engaging in Argument from Evidence

Sprawl can also eliminate entire ecosystems. When wetlands are lost, we lose many plants and animals, as well as the wetlands' ability to naturally filter water. One other problem with sprawl is the increased use of automobiles. When people live far from the center of a town or a city, they are more likely to drive to get the things they need. Increased driving leads directly to more pollution. This in itself causes many problems for the air, the water, and human health.

Not to worry, though! The news isn't all bad. Plenty of environmental problems are tied to the ways we use the land, but there are several ways we can reduce or eliminate these problems. All over the country, states, cities, and towns are working hard to keep land healthy. Let's explore a few of the methods people are using to prevent land-use problems.

One way to protect land is to require permits for all new development. For example, in many places, before people can build a house, a warehouse, or a store, they must file paperwork and pay a fee. City or county staff members then review this permit application and decide if the new construction is appropriate. If the development plan follows all of the guidelines, the project is allowed to continue. If the plan is deemed harmful, the project will be stopped.

Conservation Easement

Land Use and Planning *cont'd*

The fees that permit holders pay are meant to help offset some of the costs of construction. Permits help cities keep track of how the land is being used.

Another method to protect land and prevent problems is zoning. Zoning is the process by which certain uses for land are allowed only in certain areas. Zoning also acts as a type of land-use restriction. For example, an area that has been zoned for residential use can only include houses. Large factories cannot be built in residential zones. The zoning regulations keep the pollution and noise of industry away from neighborhoods. Zoning can also require that some areas have buildings that are built close together. This can help concentrate, rather than spread out, the impact of that land use.

One other method to prevent environmental land problems is a conservation easement. A conservation easement helps protect open space or rural land by allowing a group of people (or a whole town) to buy a piece of rural land. The land is then conserved, and it can never be developed. Many times, farmers are still allowed to farm the land, but they may not build more houses on it. This keeps that rural land open and natural, even if a nearby city grows to reach the border of the land.

Do you know how your area helps protect the land and prevent land-use issues? Can you think of any ways your town could better protect the land?

Answer the question below.

Write three statements about the passage you just read. Three of them should be true, and one should be a lie.

_____
_____
_____
_____
_____
_____
_____
_____
_____

Trade your statements with a partner. Which of your partner's statements was a lie, and why?

_____
_____
_____

Concept 4.3: Reducing Human Impacts on the Environment

CONCEPT 4.3 | How can you reduce the impact cattle have on climate change and the environment?

Activity 10
Evaluate

Responsible Land Use

Quick Code
ca6814s

Sort the following activities from having the least impact on land (1) to the most impact on land (4). **Write** your answers in the right column.

- herd grazing (moving herds over time to not overgraze)
- planting crops (using crop rotation)
- keeping as a conservation area
- logging (with tree replacement)

Least impact (1) to Most Impact (4)	Impact
1	
2	
3	
4	

CCC Cause and Effect

Explain your reasoning for the order you chose.

How Do Different Design Solutions Help Solve and Prevent Land-Use and Environmental Problems?

Activity 11
Analyze

Questions and Answers: Remote Sensing

Quick Code
ca6815s

Scan the passage that follows and identify three questions. **Write** down each question in the space below. As you read the passage, **look** for the best sentence that answers each question. Then, **add** supporting details as a bullet point list under the chosen sentence.

Questions and Answers: Remote Sensing

This passage is an interview with Dr. Nicholas Short, a renowned NASA scientist. He has published numerous articles on the technology behind remote sensing and ground truth.

Q: What is remote sensing?

A: The space program has launched many satellites into orbit around Earth. Some of these satellites look outward and gather information about space. Other satellites look inward toward Earth and gather information about Earth's surface. Both types of satellites use remote sensing. **Remote sensing** is the process of collecting information about faraway things using high-tech sensors.

Questions and Answers: Remote Sensing *cont'd*

I hear students comment that remote sensing must be concerned with remote controls, like for your home television set. But in the case of remote sensing, we're using the word *remote* in the sense of distance; *remote* means "far away."

Remote sensing provides us with data about Earth or space. We don't just use satellites. We can also use telescopes to gather images of space and aircraft to gather images of Earth.

Q: What kind of information is collected by the remote sensors?

A: Remote sensing technology collects data, usually in the form of images. If you've seen satellite photos of Earth, you've seen products of remote sensing. Often, remote sensing is used to measure how things change over time. We can measure how rivers slowly change course and how coastlines grow or shrink after a large storm. We can see the impact of a large forest fire by using "before and after" images to measure how the forest changed. The applications for this technology, especially as applied to Earth, are practically unlimited.

This image of Earth was taken using remote sensing.

Q: How can remote sensing help with making predictions about land use?

A: One of the key concepts in remote sensing is that we compare new images with previous images. This comparison can highlight how the land is changing. Our human imprint on the land is really quite shocking. I've analyzed photos that show entire swaths of rain forest cut down and ruined. I've also seen images that show how a city has grown over 20 or 30 years.

Ground truthing involves verifying remote sensing and adding data to the report.

Questions and Answers: Remote Sensing *cont'd*

From these images, we can start to recognize trends. We can quantify the information we see in images. In other words, if we notice that for each of the last 10 years, over 5,000 acres of forest were cut down in an area, we can predict the deforestation will continue at the same rate. We can then predict that in the next 10 years, we're likely to lose an additional 50,000 acres of forest in that area. We can predict how cities will grow, how resources will be used, and how humans will use land. Remote sensing is very valuable; without it, we're just guessing. With it, we're making verifiable predictions that are based on science, not opinion.

The field of remote sensing is also verified by a technique called ground truthing. This means that people visit the site of images and verify the information in the satellite photo. The people involved in ground truthing will collect additional measurements or observations about the site. For instance, a ground truthing crew might discover that the effects of sprawl are also ruining the habitats of endangered birds and insects. The ground truth provides information that might not be seen in a satellite photo. When used together, these two techniques give us more information about what is happening on Earth's surface and can ultimately help us control our negative impacts on land.

Activity 12
Observe

Green Revolution: Green Roofs

Watch the video Green Revolution: Green Roofs.

Quick Code
ca6816s

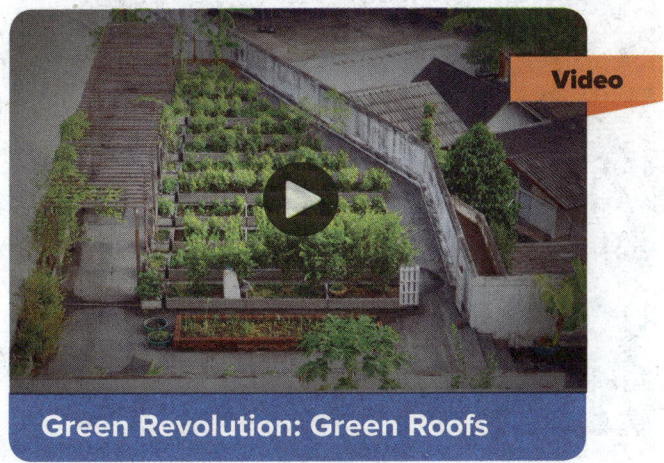

Green Revolution: Green Roofs

Write down five ideas that support the main idea of the video segment.

_____
_____
_____
_____
_____
_____
_____
_____
_____

CONCEPT 4.3 | How can you reduce the impact cattle have on climate change and the environment?

Activity 13
Observe

Change Your Diet, Save the World

Watch the video, and then **answer** the question that follows.

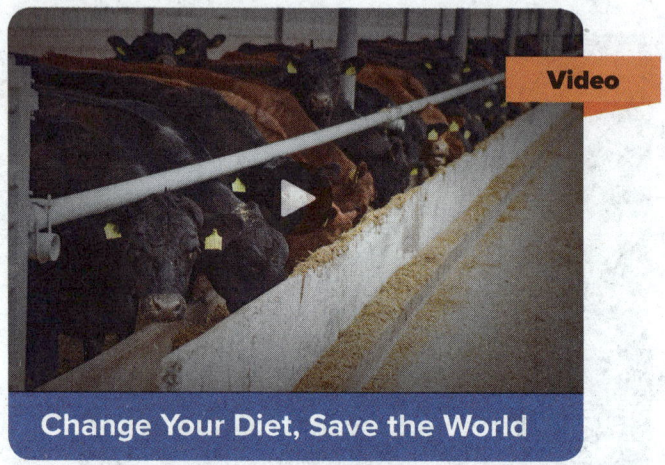

Quick Code
ca6817s

What are some possible changes that people can make to reduce their impact on the environment? Record your answers below. These ideas may come from what you have read as well as from the video. Place a checkmark next to things you can commit to do.

_____
_____
_____
_____
_____
_____
_____

Activity 14
Evaluate

Reviewing a City Plan

Quick Code
ca6818s

You have been hired as a city planner. Your first task is to review the plan for a new city. The city will be next to the foothills of mountains, with a river running through the middle of it. **Review** the provided image of the city.

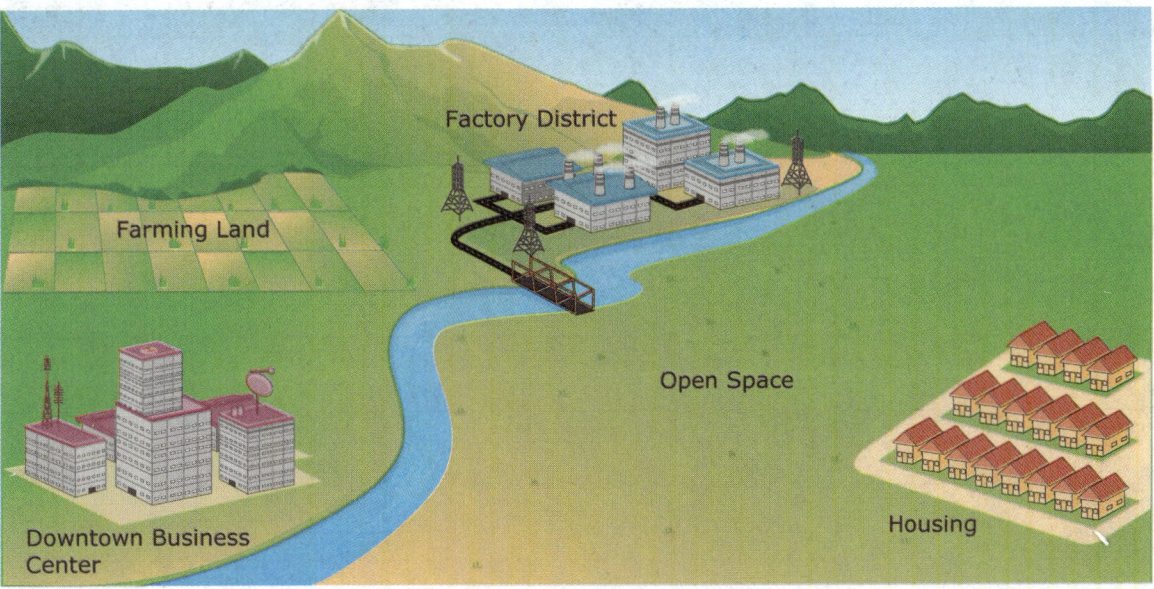

| SEP | Engaging in Argument from Evidence |

Concept 4.3: Reducing Human Impacts on the Environment

CONCEPT 4.3 | How can you reduce the impact cattle have on climate change and the environment?

Identify three or more environmental problems with the city plan.

Provide at least three solutions to your identified issues.

CONCEPT 4.3 — How can you reduce the impact cattle have on climate change and the environment?

Activity 15

Record Evidence

Cattle Population in the United States

Quick Code
ca6819s

As you worked through this lesson, you investigated and gathered evidence about reducing human impacts on the environment. Now, take another look at the Cattle Population in the United States map, which you first saw in Engage.

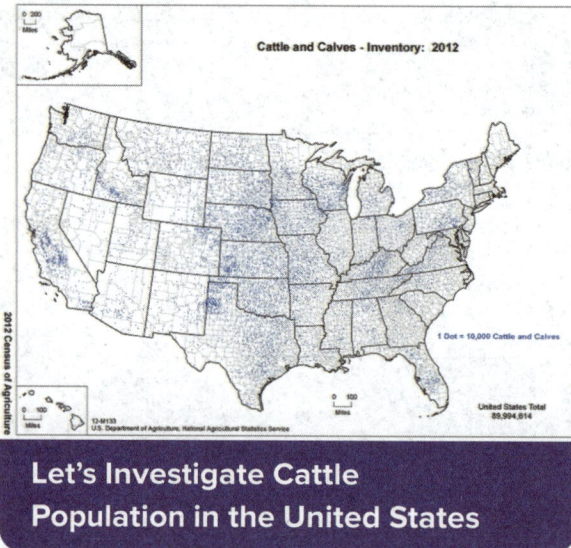

Let's Investigate Cattle Population in the United States

How has your understanding of the Cattle Population in the United States map changed?

Read the Can You Explain? question from the beginning of this lesson.

> **Can You Explain?**
>
> How can you reduce the impact cattle have on climate change and the environment?

Use your new understanding of the Cattle Population in the United States map to write a scientific explanation answering a question.

1. **Choose** a question. You may choose to answer the Can You Explain? question or one of the questions you wrote at the start of this lesson.

2. You should use a scientific explanation to answer this question. Be sure to include a claim, evidence, and reasoning.

Talk Together

Engaging in Argument: Choose a scientific explanation from one concept in the unit. Compare and critique the responses of your group members. Did they use the same evidence? Did they interpret the facts like you? How could your response have been stronger?

| SEP | Engaging in Argument from Evidence |
| SEP | Constructing Explanations and Designing Solutions |

STEM in Action

Activity 16
Analyze

How Do Engineers Reduce Human Impact on the Environment?

Quick Code ca6820s

Read the following text and **watch** the video. As you read and watch, **underline** the negative aspects of using oil and gas for fuel.

How Do Engineers Reduce Human Impact on the Environment?

Traditionally, humans have generated much of their energy by burning things. We burn wood to make fires, coal to produce electricity for our homes, and gasoline to make our cars run.

Burning fuels has its drawbacks, however. Oil and gas are located underground. Finding these resources and recovering them for our use requires drilling and transport, which can cause habitat loss. In addition, once something is burned, it often contributes to air pollution, particularly greenhouse gases.

Scientists are searching for ways to power cars without burning oil, as we seek to lessen habitat loss and reduce pollution. Some of these exciting energy innovations include electric vehicles. Automotive engineers have also designed hybrid vehicles, cars that use both gasoline and electricity for power. Drivers can charge their cars for shorter drives and use gasoline when traveling long distances. An engineer must take many factors into consideration when designing these cars, including their mass and shape. (Lighter, more aerodynamic vehicles require less fuel.) Automotive engineers are leading the way to new forms of 21st-century transportation.

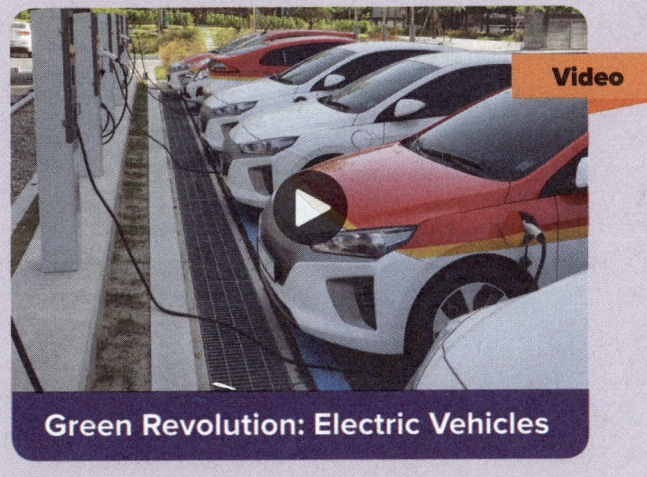

Green Revolution: Electric Vehicles

How does the electric car address the negative aspects of using oil and gas for fuel?

_____
_____
_____
_____
_____
_____
_____

Cause and Effect

What is a cause and what is an effect? **Write** either cause or effect following each impact.

air pollution _____

global warming _____

mining _____

manufacturing products _____

clearing land for farming _____

habitat loss _____

water pollution _____

burning fuel _____

Explain Your Reasoning

Pick one cause and one effect and explain how the cause could lead to the effect.

Concept 4.3: Reducing Human Impacts on the Environment

CONCEPT 4.3
How can you reduce the impact cattle have on climate change and the environment?

Activity 17
Concept Review

Review: Reducing Human Impacts on the Environment

Quick Code
ca6823s

> Now that you have completed the objectives for this concept, review the core ideas you have learned. Record some of the core ideas below.

Core Ideas

Talk with a Group

Now, think about the Measuring Climate Change map you saw in Get Started. Discuss how what you've learned about reducing human impacts on the environment can help you understand the Measuring Climate Change map.

Concept 4.3: Reducing Human Impacts on the Environment

UNIT 4 | Unit Project

Solve Problems

Unit Project: Cow Pollution

Do cows contribute to global warming? **Read** the text and **watch** the video. Then, **complete** the activities that follow.

Quick Code
ca6825s

Hereford Cow

- **SEP** Constructing Explanations and Designing Solutions
- **SEP** Obtaining, Evaluating, and Communicating Information
- **CCC** Cause and Effect

If you have ever been on a cattle ranch or dairy farm, you probably noticed that cows do a lot of belching. But did you know that cow burps contribute to global warming? Cows emit large quantities of the greenhouse gas methane. Although methane accounts for only about 10 percent of U.S. greenhouse gas emissions, it is very potent, with up to 30 times the heat-trapping power of carbon dioxide.

Methane is produced by microbes in the cows' stomach as a byproduct of the digestion process. Cow manure also releases methane as it decomposes. Estimates are as high as 400 pounds of methane per cow per year. With over a billion cows on the planet, that's a lot of methane! The problem is predicted to get worse in the future as more and more cows will be needed to meet the world's rising demand for meat.

On the bright side, methane has many important uses. It can be burned as a fuel in gas turbines and steam generators to produce electricity. It is the primary component of the natural gas that is piped into homes for heating and cooking. Most of the methane used for these purposes comes from geologic sources below ground. Could we capture the methane cows produce and use it to produce energy?

Are Cows Causing Global Warming?

Unit Project

Cows and Climate

How do cows contribute to global warming? **Select** the correct response.

○ Cows produce carbon dioxide, which is a greenhouse gas that causes global warming.

○ Cows produce methane, which blocks the sun and cools the atmosphere.

○ Cows produce carbon dioxide, which blocks the sun and cools the atmosphere.

○ Cows produce methane, which is a greenhouse gas that causes global warming.

Research Climate Impacts on Cows

Cows contribute to global warming, but climate change also affects cows. Conduct research on this topic and summarize your findings on a separate sheet of paper. Be sure to cite your sources.

Design a Solution

With your group, use the Engineering Design Sheet to create a plan to help reduce the negative impact that cows have on the environment. The solution must be something that can be implemented at an individual farm. It must be inexpensive and safe to operate. It must result in energy production for the farm or community. Draw your design on a separate sheet of paper.

Unit Project

Student Engineering Design Sheet

- What is the problem you are trying to solve?
- What are some initial ideas on how to solve the problem?
- Which of the proposed solutions deserve further investigation?

Define the Problem

Develop Solutions

- How will you construct and test your proposed solution?
- What were the results of your multiple tests?

Optimize Design

- What changes will you make to your proposed solution as a result of your tests?
- What new tests will you conduct to optimize your proposed solution?
- What is your final proposed solution to the problem?

Design Modifications

Share your design with the class and record your classmates' feedback. Then, **describe** changes you can make to improve your design.

Grade 6 Glossary

English — A — Español

adaptation
the process by which the characteristics of a species change over many generations in response to the environment (related word: adapt)

adaptación
el proceso por el cual cambian las características de una especie a lo largo de muchas generaciones en respuesta al ambiente (palabra relacionada: adaptarse)

---

air
the part of the atmosphere closest to Earth; the part of the atmosphere that organisms on Earth use for respiration

aire
parte de la atmósfera más cercana a la Tierra; la parte de la atmósfera que los organismos que habitan la Tierra utilizan para respirar

---

allele
one member of a pair or series of genes on specific chromosomes in specific positions

alelo
uno de los miembros de un par o serie de genes en cromosomas específicos en posiciones específicas

---

Antarctic
having to do with continent of Antarctica or the area around it; having to do with the South Pole or the area around it

Antártico
relativo al continente Antártico o al área que lo rodea; relativo al Polo Sur o al área que lo rodea

---

anthropogenic
caused by humans

antropogénico
causado por los humanos

Arctic
the region around the North Pole defined by the Arctic Circle, or pertaining to physical or biological features related to the region

Ártico
región alrededor del Polo Norte definida por el Círculo Polar Ártico o perteneciente a las características físicas relacionadas con la región

---

asexual reproduction
reproduction not requiring fertilization; only involves one parent

reproducción asexual
reproducción que no requiere fertilización; solamente incluye un padre o una madre

---

atmosphere
the layers of gases that surround a planet

atmósfera
capas de gases que rodean un planeta

Grade 6 Glossary

B

behavior
all of the actions and reactions of an animal
(related word: behave)

conducta
todas las acciones y reacciones de un animal
(palabra relacionada: comportarse)

biodiversity
the many different types of life that exist in an environment

biodiversidad
muchos y diferentes tipos de vida que existen en un medioambiente

biome
a major ecological community such as grassland, tropical rain forest, or desert

bioma
comunidad ecológica importante, como la pradera, la selva tropical o el desierto

biosphere
that part of Earth in which life can exist

biosfera
parte de la Tierra donde puede existir la vida

C

cardiovascular system
the body system that delivers blood to different parts of a body; consists of the heart, blood, and blood vessels
(related word: circulatory system)

sistema cardiovascular
sistema que envía sangre a distintas partes del cuerpo; incluye el corazón, la sangre y los vasos sanguíneos
(palabra relacionada: sistema circulatorio)

cell
the basic unit of all living things
(related word: cellular)

célula
unidad básica de todos los seres vivos
(palabra relacionada: celular)

cell division
the process in which a cell splits in two; part of the process in which cells reproduce

división celular
proceso por el cual una célula se divide en dos; parte del proceso por el cual se reproducen las células

cell membrane
the boundary that surrounds a cell and controls which substances can enter or leave the cell

membrana celular
contorno que rodea una célula y controla qué sustancias pueden ingresar a la célula o salir de ella

cell nucleus
an organelle in a cell that holds the cell's DNA (plural: nuclei; related term: nucleic acid)

núcleo celular
organelo de la célula que contiene el ADN de la célula (plural: núcleos; palabra relacionada: ácido nucleico)

cell wall
a stiff structure that surrounds and protects a cell; found in plant, fungus, and some bacteria cells

pared celular
estructura rígida que rodea y protege una célula; se encuentra en las células de las plantas, los hongos y algunas bacterias

cellular respiration
the process that occurs when the chemical energy of "food" molecules, carbohydrates, fats, and proteins, is released and partially captured in the form of adenosine triphosphate (ATP)

respiración celular
proceso que ocurre cuando la energía química de las moléculas de "comida", hidratos de carbono, grasas y proteínas, se libera y es capturada parcialmente en forma de trifosfato de adenosina (ATP)

Celsius
the metric temperature scale: Water boils at 100 degrees Celsius, and it freezes at 0 degrees Celsius.

Celsius
escala de la temperatura métrica: el agua hierve a los 100 grados centígrados y se congela a los 0 grados centígrados

chloroplast
an organelle in a plant cell that turns energy from the sun into chemical energy for the plant to use

cloroplasto
organelo de una célula vegetal que transforma la energía del sol en energía química para uso de la planta

circulatory system
the system that transports blood and other fluids throughout the body

sistema circulatorio
sistema que transporta la sangre y otros líquidos por el cuerpo

Grade 6 Glossary

climate
the average weather conditions in an area (related word: climatic)

clima
condiciones promedio del tiempo en un área (palabra relacionada: climático)

cloud
a collection of water droplets or ice crystals in the atmosphere

nube
colección de gotitas de agua o cristales de hielo en la atmósfera

condensation
the process by which a gas changes into a liquid

condensación
proceso mediante el cual un gas cambia a estado líquido

conduction
the transfer of heat energy within an object, or between objects that are directly touching each other, due to collisions between the particles in the objects

conducción
transferencia de energía térmica en el interior de un objeto, o entre objetos que están en contacto directo, debido a las colisiones entre las partículas de los objetos

convection (weather)
the rising of hot air above cooler air, which produces air currents

convección (tiempo)
ascenso del aire caliente por encima del aire frío, lo que produce corrientes de aire

Coriolis effect
the apparent effect of a rotating body that influences the motion of any object or fluid moving over it

efecto Coriolis
efecto aparente de un cuerpo que rota que influye en el movimiento de cualquier objeto o fluido que se mueve sobre él

current
the movement of electric charges through a material; the speed at which electric charges move through a material; the movement of liquid through a passageway or through another liquid

corriente
movimiento de cargas eléctricas a través de un material; la velocidad a la cual se mueven las cargas eléctricas a través de un material; el movimiento del líquido a través de un conducto o a través de otro líquido

D

data
information used for analysis and reason

datos
información usada para analizar y razonar

deforestation
the clearing, by burning or logging, of trees in a forested area

deforestación
desmonte, por quema o tala, de árboles en un área forestal

density
an object's mass divided by its volume; a measure of how many particles are packed together into a certain amount of space
(related word: dense)

densidad
masa de un objeto dividida por su volumen; medida que indica en qué forma se comprimen las partículas en cierta cantidad de espacio
(palabra relacionada: denso)

desert
an area that gets little precipitation and has very little vegetation

desierto
área que recibe muy poca precipitación y tiene muy poca vegetación

diaphragm
a wide muscle that separates the chest cavity from the abdominal cavity and aids in the inflation of the lungs

diafragma
músculo ancho que separa la cavidad torácica de la cavidad abdominal y ayuda a inflar y desinflar los pulmones

diffusion
dispersion of substances in a gas or liquid

difusión
dispersión de una sustancia en un gas o líquido

digestion
the process by which the body breaks down food so that it can be used for energy
(related word: digest)

digestión
proceso por el cual el cuerpo divide la comida para que pueda ser usada como energía
(palabra relacionada: digerir)

digestive system
a group of seven organs that break down food and absorb the nutrients which the body then uses for fuel

sistema digestivo
grupo de siete órganos que desintegran los alimentos y absorben los nutrientes que luego el cuerpo usa como energía

dominant trait
a genetic trait passed from parent to child that is more likely to be expressed

rasgo dominante
rasgo genético transmitido de padres a hijos que es más probable que se manifieste

Grade 6 Glossary

drought
a prolonged shortage of rainfall

sequía
falta prolongada de lluvia

--- E ---

Earth
the third planet from the sun; the planet on which we live
(related words: earthly; earth - meaning soil or dirt)

Tierra
tercer planeta desde el sol; planeta en el cual vivimos
(palabras relacionadas: terrenal; tierra en el sentido de suelo o suciedad)

ecosystem
all the living and nonliving things in an area that interact with each other

ecosistema
todas las cosas vivientes y no vivientes en un área que interactúan entre sí

egg
an animal's female reproductive cell
(related word: ovum)

huevo
célula reproductiva de una hembra animal
(palabra relacionada: óvulo)

El Niño
a band of warm ocean water temperatures that periodically develops off the Pacific coast of South America

El Niño
una corriente de aguas oceánicas de temperaturas cálidas que se desarrolla periódicamente en las cercanías de las costas del Pacífico en América del Sur

energy
the ability to do work or cause change; the ability to move an object some distance

energía
capacidad para hacer un trabajo o producir un cambio; capacidad para mover un objeto a cierta distancia

energy transfer
the transfer of energy from one organism to another through a food chain or web; or the transfer of energy from one object to another, such as heat energy

transferencia de energía
transmisión de energía desde un organismo a otro a través de una cadena o red de alimentos; o transferencia de energía desde un objeto a otro, como por ejemplo la energía del calor

environment
all the living and nonliving things that surround an organism

medio ambiente
todos los seres vivos y objetos sin vida que rodean a un organismo

equator
an imaginary line that divides Earth into Northern and Southern Hemispheres; located halfway between the North and South Poles (related word: equatorial)

ecuador
línea imaginaria que divide la Tierra en hemisferio norte y hemisferio sur; ubicada a mitad de camino entre el Polo Norte y el Polo Sur (palabra relacionada: ecuatorial)

esophagus
a muscular tube that helps move food from the mouth to the stomach

esófago
tubo muscular que ayuda a mover los alimentos de la boca al estómago

eukaryotic
having complex cells in which the genetic material is contained inside a nucleus

eucariótico
que tiene células complejas; el material genético se encuentra dentro de un núcleo

evaporation
the process in which matter changes from a liquid to a gas (related word: evaporate)

evaporación
proceso por el cual la materia cambia de estado líquido a estado gaseoso (palabra relacionada: evaporar)

excretory system
the system of the body responsible for storing and getting rid of waste products, such as urine

sistema excretor
sistema del cuerpo encargado de almacenar y de eliminar desechos, como la orina

---- F ----

F1 generation
the first generations of offspring of a genetic cross

generación F1
las primeras generaciones de descendencia de una cruza genética

fertilization
the process in which two gametes, such as an egg and sperm, unite to form a new organism, or zygote

fertilización
proceso en el cual dos gametos, como un óvulo y un espermatozoide, se unen para formar un nuevo organismo, o cigoto

Grade 6 Glossary

fossil
evidence that an organism once existed in an area; can be a part of the organism's body or a trace fossil which is a mark or print left by the organism
(related word: fossilize)

fósil
muestra de que un organismo existió una vez en un área; puede ser una parte del cuerpo del organismo o un rastro fósil, que es una marca o impresión dejada por el organismo
(palabra relacionada: fosilizar)

freshwater
water that is not salty, such as that found in streams and lakes

agua dulce
agua que no es salada, como por ejemplo la que se encuentra en arroyos y lagos

front
in weather, the boundary between two masses of air with different properties

frente
con respecto al tiempo, línea divisoria entre dos masas de aire con distintas propiedades

fuel
any material that can be used for energy

combustible
todo material que puede usarse para producir energía

function
the kind of action or activity specific to a thing or person

función
el tipo de acción o actividad específica de una cosa o persona

---------- G ----------

gametes
the reproductive cells: egg (female) or sperm (male)

gametos
células reproductivas: huevo (femenino) o esperma (masculino)

gas
a state of matter without any defined volume or shape in which atoms or molecules move about freely

gas
estado de la materia sin volumen ni forma definidos en el cual los átomos o moléculas se mueven casi libremente

gene
the basic unit of heredity in a living organism; a segment of DNA or RNA

gen
unidad básica de la herencia en un organismo vivo; segmento de ADN o ARN

generation
a group of related organisms making up a single step in the line of descent

generación
grupo de organismos relacionados que forman un solo paso en la línea de la ascendencia

genotype
internal genetic coding passed on from one generation to the next

genotipo
codificación genética interna que pasa de una generación a la próxima

geosphere
Earth's crust, both beneath the oceans and continents, as well as the mantle and inner and outer core

geosfera
corteza terrestre, tanto debajo de los océanos como de los continentes, así como también el manto y los núcleos interior y exterior

global warming
the slow warming of Earth's atmosphere due to climatic change

calentamiento global
lento calentamiento de la atmósfera de la Tierra debido al cambio climático

global wind
the atmospheric circulation around Earth that occurs in predictable patterns

vientos planetarios
circulación atmosférica alrededor de la Tierra que ocurre siguiendo patrones predecibles

gravity
a force that exists between any two objects that have mass (related word: gravitational)

gravedad
fuerza que existe entre dos objetos cualquiera que tienen masa (palabra relacionada: gravitacional)

greenhouse gas
a gas, usually carbon-based, that contributes to global warming through the greenhouse effect, which prevents the escape of radiant heat from Earth's atmosphere

gas invernadero
gas, por lo general a base de carbono, que contribuye al calentamiento global mediante el efecto invernadero, el cual impide que el calor radiante salga de la atmósfera terrestre

groundwater
water stored below Earth's surface in soil and rock layers

agua subterránea
agua almacenada por debajo de la superficie de la tierra, en capas de suelo y rocas

Grade 6 Glossary

growth
the process of using energy and dividing cells to become larger

crecimiento
proceso de uso de la energía y división de células para convertirse en uno más grande

--- **H** ---

habitat
the location in which an organism lives

hábitat
lugar donde vive un organismo

hail
small, icy balls that fall from the sky

granizo
pequeñas pelotas de hielo que caen del cielo

heat
the transfer of thermal energy

calor
transferencia de energía térmica

hemisphere
one half of a sphere

hemisferio
mitad de una esfera

heredity
the passing of traits from parent to offspring

herencia
paso de los rasgos de padres a hijos

heterozygous
having different pairs of genes for any given pair of hereditary characteristics

heterocigota
que tiene diferentes pares de genes para cualquier par determinado de características hereditarias

hibernate
to reduce body movement during the winter in an effort to conserve energy (related word: hibernation)

hibernar
reducir el movimiento del cuerpo durante el invierno con la finalidad de conservar la energía (palabra relacionada: hibernación)

homozygous
having identical pairs of genes for any given pair of hereditary characteristics

homocigota
que tiene pares idénticos de genes para cualquier par determinado de características hereditarias

hydrosphere
all of the water on, under, and above Earth

hidrósfera
toda el agua que se encuentra sobre, debajo y en la Tierra

--- I ---

ice core
a sample of ice taken by a hollow tube from a glacier or other large ice body

núcleo de hielo
muestra de hielo que se toma de un glaciar o de otro gran cuerpo de hielo mediante un tubo hueco

inherit
to receive genetic information and traits from a parent or parents (related word: inheritance)

heredar
recibir información y rasgos genéticos de un padre o de los padres (palabra relacionada: herencia)

instinct
an animal's natural response to a stimulus; an inherited behavior

instinto
respuesta natural de un animal a un estímulo; comportamiento heredado

--- J ---

joint
the place where two bones are connected in a body

articulación
lugar en que dos huesos se unen en el cuerpo

--- K ---

kinetic energy
the energy an object has due to its motion

energía cinética
energía que tiene un objeto debido a su movimiento

--- L ---

larynx
a tube above and continuous with the trachea; contains the structures used for speech; also called the voice box.

laringe
tubo inmediatamente continuo a la tráquea, contiene los órganos encargados de producir los sonidos del habla u órganos fonadores

Glossary | Grade 6 | R11

Grade 6 Glossary

latitude
angular distance north and south of the equator

latitud
distancia angular al norte y sur del ecuador

light energy
that form of energy that animals can see directly; visible electromagnetic radiation

energía lumínica
tipo de energía que los animales pueden ver directamente; radiación electromagnética visible

liquid
a state of matter with a defined volume but no defined shape and whose molecules roll past each other

líquido
estado de la materia con un volumen definido pero no forma definida y cuyas moléculas se deslizan unas sobre otras

lungs
organs of the respiratory system that bring oxygen-rich air into the body and send oxygen-poor air out of the body

pulmones
órganos del sistema respiratorio que traen aire rico en oxígeno al cuerpo y envían aire pobre en oxígeno fuera del cuerpo

M

meteorite
a piece of rock or metal from space that strikes Earth's surface

meteorito
pieza de roca o metal del espacio que golpea la superficie de la Tierra

migration
the movement of a group of organisms from one place to another, usually due to a change in seasons

migración
movimiento de un grupo de organismos de un lugar a otro, generalmente debido a un cambio de estaciones

mitochondria
an organelle in eukaryotic cells that is the site of cellular respiration and generates most of the cell's ATP

mitocondria
orgánulo de las células eucariotas donde tiene lugar la respiración celular y donde se genera la mayor parte del ATP de la célula

model
a simulation of a real thing or process

modelo
una simulación de una cosa o proceso real

moisture
a measure of how much water is in something (related word: moist)

humedad
una medida de cuánta agua hay en algo (palabra relacionada: húmedo)

multicellular
consisting of many cells

multicelular
que tiene muchas células

muscular system
the body system that permits movement and locomotion in animals

sistema muscular
sistema corporal que permite el movimiento y la locomoción de los animales

--- N ---

nervous system
the system of the body that carries information to all parts of the body: The nervous system relies on nerve cells to move electrical signals to the body from the brain, and from the body to the brain and/or spinal cord.

sistema nervioso
sistema del cuerpo que transporta información a todas las partes del cuerpo: El sistema nervioso depende de las células nerviosas para transportar señales eléctricas al cuerpo desde al cerebro, y desde el cuerpo al cerebro y/o la médula espinal

nitrogen
an element that makes up most of the air near Earth's surface: nitrogen is a gas at room temperature

nitrógeno
elemento que forma la mayor parte del aire que se encuentra cerca de la superficie de la Tierra: el nitrógeno es un gas a temperatura ambiente

--- O ---

ocean current
large-scale movement of water within the oceans in a certain direction

corriente oceánica
movimiento de agua en gran escala que tiene lugar dentro de los océanos en una dirección determinada

offspring
a new organism that is the product of reproduction

descendencia
organismo nuevo que es el producto de la reproducción

Grade 6 Glossary

organelle
a tiny structure within a cell that performs a specific function for that cell

organelos
estructura diminuta que se encuentra dentro de una célula y que realiza una función específica para esa célula

---- P ----

parent
an organism from which younger organisms are obtained

progenitor
organismo del que se obtienen organismos más jóvenes

pharynx
tube at the back of the mouth that leads to either the respiratory or digestive systems; also called the throat.

faringe
tubo que conecta el final de la boca con los sistemas respiratorio y digestivo; también se denomina garganta

phenotype
the observable traits of an organism passed on from parent to offspring

fenotipo
rasgos observables de un organismo que los progenitores transmiten a su descendencia

polar
in earth science, having to do with the areas on Earth closest to the geographic North or South poles; in chemistry, describes a molecule that has a positively charged side and a negatively charged side (related words: pole, polarity)

polar
en la ciencia de la tierra, tiene que ver con las áreas de la Tierra más próximas a los polos geográficos Norte y Sur; en química, describe una molécula que tiene un lado con carga positiva y otro con carga negativa (palabra relacionada: polo, polaridad)

pollination
the transfer of pollen from the stamen (male) to the stigma (female)

polinización
transferencia de polen desde el estambre (masculino) al estigma (femenino)

pollutant
undesirable or harmful substance that alters the natural balance of an ecosystem, particularly found in air, soil, or water

contaminante
sustancia indeseable o dañina que altera el equilibrio natural de un ecosistema, en particular se encuentra en el aire, el suelo o el agua

precipitation
water that is released from clouds in the sky; includes rain, snow, sleet, hail, and freezing rain

precipitación
agua liberada de las nubes en el cielo; incluye la lluvia, la nieve, la aguanieve, el granizo y la lluvia congelada

predator
an animal that hunts and eats another animal

depredador
animal que caza y come a otro animal

pressure
the force per unit area exerted on an object

presión
fuerza por unidad de área ejercida sobre un objeto

prey
an animal that is hunted and eaten by another animal

presa
animal que es cazado y comido por otro

pulmonary artery
the large artery that brings blood from the heart to the lungs

arteria pulmonar
arteria grande que lleva sangre del corazón a los pulmones

Punnett square
a chart used to determine the possible genetic outcomes for offspring of a given cross

cuadro de Punnett
gráfico utilizado para determinar los posibles resultados genéticos en la descendencia de un cruzamiento dado

R

radiation
a process by which energetic electromagnetic waves move from one place to another

radiación
proceso por el cual las ondas energéticas electromagnéticas se desplazan de un lugar a otro

rain
liquid water that falls from the sky

lluvia
agua líquida que cae desde el cielo

rain forest
a forest that is humid and rainy for much of a year

selva tropical
bosque húmedo y lluvioso la mayor parte del año

Grade 6 **Glossary**

recessive
a genetic trait that lacks the ability to manifest itself when a dominant gene is present

recesivo
rasgo genético que carece de la posibilidad de manifestarse cuando está presente un gen dominante

remote sensing
making observations of a body without making physical contact

detección remota
hacer observaciones de un cuerpo sin hacer contacto físico

reproduce
to make more of a species; to have offspring (related word: reproduction)

reproducir
hacer más de una especie; tener descendencia (palabra relacionada: reproducción)

respiratory system
the system of the body that brings oxygen into the body and releases carbon dioxide

sistema respiratorio
conjunto de órganos del cuerpo que hace ingresar oxígeno al cuerpo y libera dióxido de carbono

─── **S** ───

salinity
the total quantity of dissolved salts in water

salinidad
cantidad total de sales disueltas en agua

seed
a plant structure that contains a young plant, food supply, and protective coating

semilla
estructura de una planta que contiene una planta joven, reservas de alimentos y una capa protectora

sexual reproduction
a biological process by which plants and animals create offspring by combining their genetic material

reproducción sexual
proceso biológico por el cual las plantas y los animales crean descendencia mediante la combinación de su material genético

skeletal system
the network of solid materials that give an organism's body its structure

sistema esquelético
red de materiales sólidos que proporcionan al cuerpo de un organismo sus estructuras

solar energy
energy that comes from the sun

energía solar
energia que proviene del Sol

solid
matter with a fixed volume and shape

sólido
materia con un volumen y una forma determinada

sperm
the male gamete; a reproductive cell that can fertilize an egg cell

espermatozoide
gameto masculino, célula reproductiva masculina que puede fertilizar una célula óvulo

stimulus
something that prompts a change in an organism's behavior (related word: stimuli)

estímulo
algo que motiva un cambio en la conducta de un organismo (palabra relacionada: estímulos)

subsystem
a working system that is part of a larger system

subsistema
un sistema de trabajo que es parte de un sistema más grande

sunspot
a cooler darker spot on the surface of the sun

mancha solar
mancha más oscura y fría sobre la superficie del Sol

system
a group of parts that work together to function or perform a task

sistema
un grupo de partes que trabajan juntas para funcionar o realizar una tarea

---
T
---

taiga
a large naturally occurring area of land with largely evergreen forest vegetation found in northern sections of the Northern Hemisphere

taiga
área natural de tierra con grandes bosques perennes que se encuentra en secciones norteñas del hemisferio norte

temperate
describes a climate that does not have extremes of hot or cold

templado
describe un clima que no tiene extremos de calor o frío

Grade 6 **Glossary**

temperature
a measure of the average kinetic energy of the atoms in a system, used to express thermal energy in degrees

temperatura
medida del porcentaje de energía cinética de los átomos de un sistema, se usa para expresar la energía térmica en grados

thermal energy
energy in the form of heat

energía térmica
energía en forma de calor

topography
detailed mapping of the physical features of a small locale or area

topografía
características físicas que definen el relieve de un lugar, como montañas, valles y la forma de los accidentes geográficos

tropical
describing a climate that is very hot and humid; describing an area of Earth that is near the equator

tropical
describe un clima que es muy cálido y húmedo; describe un área de la Tierra que está cerca del ecuador

tundra
extremely cold climate located near the North and South Poles and on the tops of mountains; receives very little precipitation and has no trees

tundra
clima extremadamente frío ubicado cerca de los polos Norte y Sur y en la cima de las montañas; recibe muy pocas precipitaciones y no tiene árboles

water
a compound made of hydrogen and oxygen

agua
compuesto formado por hidrógeno y oxígeno

water cycle
the continual movement of water between the land, ocean, and the air

ciclo del agua
movimiento continuo del agua entre la tierra, los océanos y el aire

water vapor
the gaseous form of water; produced when water evaporates

vapor de agua
estado gaseoso del agua; se produce cuando el agua se evapora

weather
the properties of the atmosphere at a given time and location, including temperature, air movement and precipitation

tiempo atmosférico
propiedades de la atmósfera en un determinado momento y lugar; entre ellas, la temperatura, el movimiento de aire y las precipitaciones

wind
the movement of air due to atmospheric pressure differences

viento
movimiento de aire que se produce por las diferencias en la presión atmosférica

Index

A

Adaptation 82, 128
Allele 174, 179
Analyze 24–26, 31–36, 44–48, 58–61, 69–72, 80–81, 84, 91–97, 102–105, 118–119, 126–132, 134–137, 150–156, 169–172, 174–176, 179–182, 190–194, 222–225, 236–239, 242–245, 255–258, 266–269, 292–294, 296–297, 302–304, 310–313, 331–333, 336–337, 342–345, 348–352, 360–363
Antarctic 26
Anthropogenic 243
Arctic tundra 69, 92
Asexual reproduction 118–119, 151
Ask Questions 10–11, 68, 112–113, 162–163, 214–215, 276–277, 320–323
Atmosphere 224, 243, 255–257, 267–268, 294, 331

B

Behavior 127, 294–297, 302–305
Biodiversity 336–341
Biome 69

C

Can You Explain? 8, 54–57, 66, 100–101, 110, 146–149, 160, 188–189, 212, 262–265, 274, 306–309, 318, 358–359
Climate
    changes In 222–225, 237, 241–243, 253, 255–259, 311
    other organisms impacted by 267–268, 293–295, 297, 302–303, 369
    zones 24–26, 28, 31–36, 44–51, 58
Conduction 46
Convection (weather) 31–33
Current 34, 45, 51, 58

D

Data 242, 251, 255–257
Deforestation 294, 336, 352
Density 32, 38, 44–45
Deserts 10, 33, 54,
Design Solutions 284–289
Dominant trait 164–167, 175, 185, 187
Drought 223

E

Ecoystem 84, 92
Egg 119
El Nino 223
Energy transfer 31–32
Environment
    changes in 293–294, 397
    human impact on 331–333, 342–344, 360–361
Equator 26, 32–33, 37, 45
Evaluate 16–19, 27–28, 37, 51–52, 73–75, 82–83, 90, 98–99, 114–117, 124–125, 138–141, 164–167, 173, 185–187, 219–221, 240–241, 253–254, 261, 278–281, 295, 305, 324–325, 334–335, 346–347, 355–357

F

F1 generation 170
Fertilization 119, 127
Fossil 243
Fuel 243, 360

G

Gametes 119, 126–127
Gene 171, 174, 180–181, 192, 202
Generation 169
Genotype 178–179, 184
Global warming 243, 255–257, 268, 294, 302–303
Global wind 32–34, 37
Greenhouse gases 243, 251, 256, 294, 360, 367

H

Habitats 293–295, 302–303, 336, 360–361
Hands-On Activities 20–23, 38–43, 230–233, 284–289
Heat 24, 31–32, 46
Hemisphere 236
Heredity 164, 168–169, 196
Heterozygous 175, 177
Hibernation 297–298, 302–304
Homozygous 175

I

Ice core 255–257
Inherit 171, 175
Instinct 127
Interpret Data 87–88, 246–249, 251–252, 259–260, 300–301
Investigate 20–23, 38–43, 230–233

K

Kinetic energy 46

L

Latitude 26, 31–34, 297, 300
Light energy 31

M

Meteorite 224–225
Migration 297–299, 302, 310–313

O

Observe 12–15, 29–30, 49–50, 85–86, 89, 133, 168, 177–178, 183–184, 216–218, 226–229, 234–235, 250, 282–283, 290–291, 298–299, 353–354
Ocean current 45, 58–59
Offspring 91, 118–119, 169, 175, 190

P

Parent 168–171, 174
Phenotype 179, 184
Polar 26, 37, 46
Pollination 133–134, 138, 142, 173
Pollution 294, 331–333, 343–345, 360–363
Predator 297
Prey 297
Punnett square 174–177, 183–187

R

Radiation 31–32, 46, 222–224, 236
Rain forest 16

Index

Reason 76–79, 120–123, 326–330, 338–341
Recessive 179, 179, 185–187
Record Evidence 54–57, 100–101, 146–149, 188–189, 262–265, 306–309, 358–359
Remote sensing 348–352
Reproduce 114, 118
Reproduction 120, 124–128, 151

S

Salinity 45–46
Seed 151, 173
Sexual reproduction 119, 126–133, 151
Solar energy 20, 25–28, 46
Solve Problems 4–5, 142–145, 198–230, 208–209, 366–371
Sperm 119
STEM in Action 58–61, 102–105, 150–154, 190–194, 266–269, 310–313, 360–363
Stimuli 297, 300
Sunspot 223, 240, 242

T

Taiga 80
Temperate 16, 26–28
Tropical 16, 26–28
Tundra 92

U

Unit Project 4–5, 198–203, 208–209, 366–371

W

Water cycle 34
Water vapor 33
Weather 222, 257, 294–297
Wind 31–33, 37, 45, 58